TRAVELLING AROUND THE HUMAN GENOME

An *in situ* investigation

Bertrand Jordan
Research Director – CIML INSERM/CNRS – Marseille

Les Éditions INSERM
101, rue de Tolbiac
75654 Paris Cedex 13, France
Tel. : 33 1 44 23 60 82

John Libbey Eurotext
6, rue Blanche
92120 Montrouge, France
Tel. : 33 1 47 35 85 52

© INSERM John Libbey Eurotext, 1993
ISBN INSERM 2-85598-572-2
ISBN John Libbey Eurotext 2-7420-0030-5

Acknowledgements

This book is dedicated to all the people who made its publication possible – first and foremost to my partner in life, Anne, who travelled around the world with me on this trip and was forever busy checking, tidying up, filing, correcting and, most important of all, criticizing my successive drafts. Without her this book would contain many more flaws.

A word of thanks is also due to those who believed enough in this quest to fund it – Claude Paoletti (CNRS), Bernard Barataud (AFM), not forgetting the French Ministry of Research and INSERM. Another decisive factor was the support of Axel Kahn and of *médecine/sciences*, which published my monthly "Chroniques génomiques" ("Tales of the Genome").

I would also like to acknowledge the help received from the numerous fellow scientists who welcomed me into their laboratories and to thank those generous souls who allowed me to set up my "base camp" on their home ground: Charles Cantor in the USA, Michio Oishi and Kenichi Matsubara in Japan, and Walter Bodmer in Great Britain. I am also indebted to Corinne Beziers La Fosse for most of the artwork and to the administration of the Immunology Centre at Marseille – in particular to Daniel Francal (administrator) and Marylise Issa (Head of Secretarial Services) – for their faraway but often invaluable assistance.

This study has grown into a book thanks to the backing of Editions INSERM, notably in the person of Suzy Mouchet. The English version was kindly reviewed by Bob Cook-Deegan, who had provided me with a prepublication draft of his extensive study, *Gene Quest*, on the science and politics of the Human Genome project [*]; remaining errors are of course solely my responsibility. Last of all, I must acknowledge the developers of the notebook computer, a fantastic tool without which this work would have been immensely more difficult.

---

[*] **Cook-Deegan RM:** *The Gene Wars: Science, Politics and the Human Genome.* **Norton,** New York, 1993.

Table of contents

1. The birth of an investigation 1
Human genetics in the DNA era 1
Reverse genetics: major progress and limits 1
An unconventional sabbatical 2

2. Emerging Human Genome Programmes 5
Form and content 5
Two types of maps: genetic and physical 6
The gradual refinement of genetic maps 8
Physical maps make headway 8
Sequencing: methodology and cDNA 11
National policies
    USA... 12
    Japan... 13
    Europe... 13
    The EEC... 15
    ... and HUGO 15

3. Families, RFLPs and microsatellites: the tools for genetic mapping . 17
A map that needs refining 17
Disappointingly slow progress 19
Motivation, recognition, funding 22
Polymorphism: a question of degree 23
Microsatellites, *alias* CA repeats 24
Other innovations 27
On the way to a two centiMorgan map 28

4. Evolving tools for physical mapping 33
Complementary methods: *in situ*, Southern blots and cloning 33
A problem of scale 35
Megabase methods:
    Pulsed field gel electrophoresis for analysis 36
    ... and YACs for cloning 36

YACs are not alone 46
A myriad of endeavours 47

5. Strategies for an integrated map 51

Physical maps come in several varieties 51
From region... to chromosome 53
Chromosome 19 at Lawrence Livermore 54
Have sequence tagged sites won the day ? 55
    The concept 55
    Where are we today ? 56
    STS, EST, microsatellites... Towards the perfect landmark ? 57
Another approach: reference libraries 58
Simultaneous YAC contig building over the whole genome:
a surprise from Généthon 60

6. Genome programmes and medical genetics: the case of the fragile X syndrome 63

Back to « reverse genetics » 63
The contribution of Genome programmes 65
The long march to the fragile site on chromosome X 66
    An enigmatic syndrome 66
    "Fishing for probes" 67
    Steve Warren's translocations and the importance of "panels" 68
    YACs come to the rescue 70
    Hypermethylation and instability 71
    Fragile X begins to be understood 72
Take-home lessons 73

7. Getting down to sequencing ? 77

Widespread optimism 77
A standard strategy 78
Disillusionment sets in 78
Technology is not the only problem 79
A fresh start on a more reasonable basis 80
An ambitious but realistic sequencing project: the Nematode 81
Yeast chromosome III: a successful consortium 82
Craig Venter's interests extend beyond cDNA 83
Genome sequencing comes of age 84
cDNA studies are very popular 84
    An effective approach 84
    The sequence as a "signature" 85

Enemy brothers ? 87
A fight over patents 87
Exotic techniques 88
Tunnelling microscopy: a hope for DNA sequencing? 90
    Scanning tunnelling microscopy and related techniques 90
    DNA sequencing by scanning tunnelling microscopy ? 90
    High quality DNA imaging 91
    What remains to be accomplished ? 92
    Sequences, DNA-binding factors and interactions 93
"Small" laboratories should not be overlooked 94

8. Genome and informatics: the infernal twins 97

Multiple requirements: data entry and interpretation 97
Computerized laboratory notebooks... 98
    An example: the Lawrence Livermore Genome Center 98
Semi-private or semi-public databases... 99
General-purpose banks... 99
... and political stakes ! 100
Bicultural coexistence is difficult 102

9. The end of "cottage industry" instrumentation ? 103

Research and manual labour 103
Automation, an obvious solution ? 104
The HUGA "sequencing factory" in Tsukuba, ... 105
... and the Généthon Mark II room 108
Some (initial) lessons from these experiences 109

10. Genome research in the top two: from Livermore to Tsukuba 111

A very exclusive club 111
In the USA: a robust genome Programme 112
    The sociology of "Genome Centers" 112
    Lessons learnt: maps are feasible 113
    The USA have taken up cDNAs... 114
    Data processing is taken seriously... 115
    The overall track record is positive ! 115
Genome in Japan: myth and reality 116
    Illusions and delusions 116
    The reality in Japan 117
    Things get underway 118
    Japan: a power to be reckoned with 121

11. The Old World is still in the race 125

Great Britain: a bread and butter genome Programme 125
   Great Britain: fertile ground 125
   The Human Gene Mapping Project: unpretentious and pragmatic 125
   The Resource Centre: a services pool for laboratories 126
   A dense and dynamic research fabric 127
   Hans Lehrach (ICRF, London) and his "reference libraries" 127
   Cosmids in England: the Nematode... 128
   ... and chromosome 11 129
   Edinburgh prepares for the post-gene era 129
   "Rule Britannia" ? 130
France: a large potential and a complex situation 131
   An honourable third place 131
   A powerful associative sector 132
   Antinomical methods of organization 133
   Is a French genome Programme necessary ? 135
   "GIP Genome" and "GREG": a drawn-out pregnancy 135
Continental Europe: the genome archipelago 137
   "Peninsularities" 137
   Germany: definitely reserved 139
   The Netherlands: Peter Pearson's legacy 140
   Scandinavia: getting organized 140
   The European Programme comes out of limbo 141

12. Coordination or competition ? 145

Rivalry is the rule 145
HUGO: necessary... but still searching for its identity 147
   A blurred public image 147
   EMBO and HUGO: very different initial conditions 147
   HUGO: a difficult childhood 148
   The stakes involved 149
   The "clout" of the United States 151
   Hopes for the future 152

13. Genome, the trouble-maker 155

A focus of misunderstandings 155
   "Sequencing man" 155
   Give a dog a bad name... and hang him 156
A mind-numbing routine ? 156
   One more misunderstanding 156
   How qualified should the staff be ? 157

The trials and tribulations of chromosome-specific libraries	158
An astute blend	159
Automation and Resource Centres	160
Généthon: research or service ?	161
Is it worth the money ?	162
The emergence of economic issues	163
« Big Science » in biology ?	163

14. The ugly ethician — 167

Beware of ethics: trespassers will be prosecuted !	167
Fallout from genome programmes	167
Poorly controlled repercussions	168
Health care systems: private enterprise or state control ?	168
Real-life experience of ethics	169
Possibly damaging catch-words	170
The benefits of contact with the patients	170
A cut for ethics	171
The melting pot of science	172
Society's outlook on research	172
France: an almost suspect indulgence	172
In Germany: a witch hunt !	173
The duty of communication	174
"Just let us get on with our work"	174
A strong demand	175
The role of research scientists	176

Postface 179

Abbreviations 181

Subject index 183

Author index 187

1
The birth of an investigation

Human genetics in the DNA era

The era of molecular biology was ushered in by Avery and McLeod [1] who demonstrated the role of DNA as the vector of heredity in 1944, and later by Watson and Crick [13] who determined its structure in 1953. Although key advances such as deciphering the genetic code were made in the sixties, the real breakthrough occurred in the seventies with the development of genetic engineering methods that led to a quantum jump in our knowledge of genes, of their expression and their regulation. In just a few years these techniques became sophisticated enough to tackle the most complex of genomes – our own. Beta-globin [12] was the first human gene to be cloned, quickly followed by many others. Shortly afterwards, clinical application of these methods began with the first prenatal diagnoses based on DNA analysis [10].

This penetration of genetic engineering into medicine was soon to move ahead at breakneck speed with the ever-widening applicability of DNA analysis for diagnosing hereditary diseases and with growing prospects for gene therapy – although difficulties were underestimated in the latter case: what had appeared close at hand in 1985 only came to fruition in 1990 [9]. At the same time the power of genetic analysis was boosted many-fold by the introduction of new "characters", i.e. DNA sequence polymorphisms called RFLPs (Restriction Fragment Length Polymorphisms) [2]. Suddenly a complete map of the human genome appeared to be within our reach.

Reverse genetics: major progress and limits

In conjunction with increasingly sophisticated techniques, this knowledge was applied to investigate diseases whose hereditary origin had been proven but for which the type of defect, the protein responsible and the pathogenic mechanism were completely unknown. At the time, this kind of study was called reverse genetics [6] because it reversed the conventional process, which sets out from the biochemical defect in search for the gene - as for hemophilia. The name is in fact rather unfortunate, since it invites confusion with a different kind of "reverse genetics", the approach in which a gene is modified *in vitro* and then reintroduced in the cell or the animal to find out the new phenotype. A more adequate designation, "positional cloning", has been advocated, but has not really caught on outside specialized circles. Whatever the name, reverse genetics recorded some brilliant successes with the discovery of the gene responsible for Duchenne's muscular dystrophy [3, 11], the gene involved in cystic fibrosis [4, 5] and many others.

However in the course of this research, some of its limits became apparent. Work took place in a series of uncoordinated programmes, each narrowly focused on a particular disease and chromosome region, and often performed simultaneously by a multitude of fiercely competing teams. This resulted in progress by fits and starts, with some parts of the genome intensively studied and others left virtually untouched - a definite waste of effort. These reasons, with many others, prompted some scientists to propose setting up a "Human Genome Programme" [7], aimed at coordinating the systematic molecular analysis of human DNA. This proposal stimulated a great deal of discussion, largely caused - as will be discussed later - by a number of misunderstandings as to its aims. Today the Human Genome Programme exists not only in the country where it was first proposed - the USA - but also in Great Britain, France [8], Italy, Japan and elsewhere. It is well funded, at least as biological research goes, and has already produced significant results. The first complete physical maps of certain chromosomes have been presented, and already the DNA sequences characterized in this work are proving extremely valuable.

An unconventional sabbatical

Such was the context in which I decided to devote one year to investigating the state of the art in this field. As a molecular biologist already involved in human genetics, such a subject appeared very tempting for a sabbatical which I had been preparing for a long time. In our research community, such a period is not usually spent indulging in a personal taste for music, for painting or for any similar pursuit. It is, in fact, quite easily granted (with full pay) by the agencies which provide the salaries of most scientists in my country, INSERM (French National Institute for Health and Medical Research) or CNRS (National Centre for Scientific Research);

it is not even necessary to wait seven years for the next sabbatical – which, however, has to be definitely science-oriented. French scientists usually spend their sabbatical year in a laboratory – as a general rule abroad – in order to update their expertise, to learn new techniques and to get back to "hands-on" research work, thanks to a temporary break from administrative work and because INSERM, CNRS, the evaluation committees and France are all far away.

My own project proposal was somewhat unusual: it consisted of a year-long study organized around visits to the hundred or so centres working on the human genome around the world. I was convinced that an in-depth, on-the-spot survey of this kind conducted over a year by a researcher active in the field would bring in a rich harvest both for me and for the French scientific authorities. As it turned out, I had virtually no problem making these see the light and I therefore set out on my trip "around the world through 80 labs" with the combined blessings of my employer (CNRS), the French Muscular Dystrophy Association (AFM), the French Ministry of Research and INSERM. No acknowledgement of my sponsors would be complete without thanking *médecine/sciences* for publishing a monthly news feature ("Chroniques Génomiques", i.e. "Tales of the Genome") in which I could freely discuss all aspects of this important project, including technical concerns, scientific policy (sometimes a touchy subject!) and even politics. This book is based on these regular features whose publication started in 1990 (as it were to whet the reader's appetite) and is still continuing today (as a well established routine). Reorganized, updated but still recognizable, these papers form the framework of this book. They were written by a French author for a French audience, and I have, in this English version, kept some of their Gallic flavour. For the same reason, some of the cited articles – often from *médecine/sciences* – are in French. This work is aimed, as are the "Tales of the Genome", at a relatively wide community of biologists, medical doctors and others interested in the topic. I can only hope that it will prove of use, and hopefully enjoyable, though certainly demanding, to read.

REFERENCES

1. Avery OT, McLeod CM, McCarty M: Induction of transformation by a desoxyribonucleic acid fraction isolated from pneumococcus type III. *J Exp Med* 1944, **79:** 137-158
2. Bostein D, White RL, Skolnick M, Davis RW: Construction of a genetic linkage map in man using restriction fragment length polymorphisms. *Am J Hum Genet* 1980, **32:** 314-330
3. Chelly J: La myopathie de Duchenne: du gène DMD à la dystrophine. *Médecine/Sciences* 1988 **4:** 141-144
4. Cox K, Chakravarti A, Buchwald M, Tsui L-C: Identification of the cystic fibrosis gene: genetic analysis. *Science,* 1989 **245:** 1073-1080
5. Goossens M: La découverte du gène de la mucoviscidose. *Médecine/Sciences* 1989 **5:** 589-591

6. Jordan B: Grandeur et servitudes de la génétique inverse.   *Médecine/Sciences* 1988 **4:** 138-140
7. Jordan B: Les cartes du génome humain.   *La Recherche* 1989 **20:** 1486-1494
8. Jordan B: Programmes Génome: et la France?   *Médecine/Sciences* 1990 **6:** 807-809
9. Kahn A: Nouvelles orientations pour la thérapie génique.   *Médecine/Sciences* 1990 **6:** 144-149
10. Kan YW, Dozy AM: Polymorphism of DNA sequence adjacent to human beta-globin structural gene: relationship to sickle mutation.   *Proc Natl Acad Sci* (USA) 1978 **75:** 5631-5635
11. Koenig M, Hoffman EP, Bertelson CJ, Monaco AP, Feener C, Kunkel LM: Complete cloning of the Duchenne Muscular Dystrophy (DMD) cDNA and preliminary genomic organisation of the DMD gene in normal and affected individuals.   *Cell* 1987 **50:** 509-517
12. Rabbits TH: Bacterial cloning of plasmids carrying copies of rabbit globin messenger RNA.   *Nature* 1976 **260:** 221-225
13. Watson JD, Crick FHC: Genetical implications of the structure of deoxyribonucleic acid.   *Nature* 1953 **171:** 737-738

2
Emerging Human Genome Programmes

Form and content

The actual launch of a Genome Programme in the USA was preceded by many discussions, which began with a meeting in Santa Cruz in May 1985 [12]. A report produced for the US Congress by the Office for Technological Assessment provided a well-balanced and clear picture of the prospects in April 1988 [11]. The programme actually got underway in 1989 with a funding of $100 million US and quickly produced scientific and technological spin-off [9], even in fields that at first sight seemed to have little in common with human genetics, such as plant biology [6].

Let us first set the stage by giving an overview of the work undertaken: contrary to frequent reports, it is not focused on all-out DNA sequencing. It is true that, at the outset, the project had been presented and debated in these terms; the announced objective was to determine the complete sequence of the human genome, i.e. something like 3,000 million bases. The ensuing academic discussions centred on whether it would be best to "sequence" a man or a woman (a man, to obtain data on the Y chromosome), whether a sequencing cost of 10 cents/base or rather $1/base was appropriate and whether specialized genome centres should be set up. With today's hindsight, these questions were obviously premature. In fact most of the studies funded by the various genome programmes come under three headings: genetic mapping, physical mapping and relatively focused sequencing work. Clearly, these three types are inter-related but the classification is convenient for analyzing them.

Two types of maps: genetic and physical

The purpose of genetic mapping [14] is to determine the relative positions of a number of hereditary characters. Based on character transmission analysis, genetic mapping consists of searching for correlations. If, when families are analyzed, two characters are found to be nearly always inherited together, it is deduced that the two respective genes are located "near" each other on the same chromosome. This analysis is based on two prerequisites: polymorphism, i.e. the existence of different "versions" of a given character (and hence of its determining gene), and accessibility to a wide set of large families with clearly defined kinships. In these conditions a degree of proximity, a distance and by extension an order can be determined. In other words, assuming the three characters A, B and C under study are "related" (relatively close to each other on the same chromosome), this method will be able to determine their order (A C B for example). This approach is only feasible when the characters being studied are determined by a single gene (i.e. "monogenic"). Characteristics such as height and intelligence depend on the interaction of many genes (not forgetting the possibly overwhelming influence of the environment !) and cannot be directly analyzed by this method.

In contrast, the physical map provides a much more detailed description of the molecular reality: it refers explicitly to the DNA molecule, which is the material support of heredity. Now the aim is to determine the position of the genes on the chromosomes, the distance, i.e. the number of nucleotides separating them along the DNA fibre. The ultimate physical map is the full nucleotide sequence of the region investigated, but much faster methods produce coarser yet useful diagrams.

Both maps can use the same landmarks; this is, in fact, highly desirable even though it is not always the case (Figure 2-1). The genetic map may include characters whose genes are not identified as long as they are polymorphic. In contrast, the physical map includes DNA regions common to all individuals but not distinguishable by genetics because they are non-polymorphic. In both maps, the landmarks will be in the same order, i.e. if character C is genetically located between A and B, this is because the respective genes coding for A, C and B occur in this order along the length of the DNA molecule. It is nevertheless perfectly possible for C to be closer to A than to B on the genetic map whereas the opposite is true on the physical map. In fact these maps are complementary and many laboratories are working on both aspects at the same time.

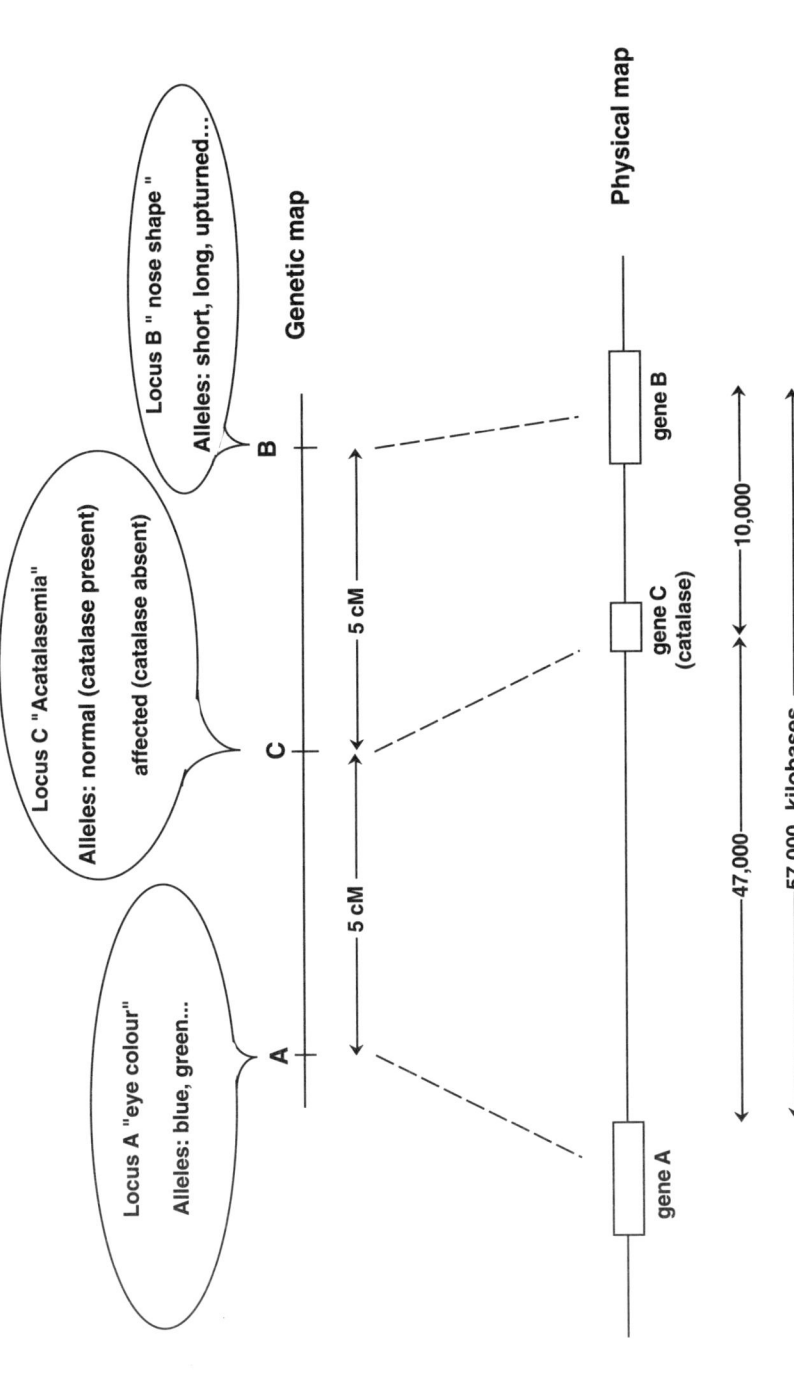

Figure 2-1 Genetic and physical maps. In this highly schematic diagram, the genetic map of three loci is outlined, with the physical map of the three corresponding genes shown underneath. The top map is derived by tracking the transmission of characters in families, followed by statistical analysis. Assuming there are DNA probes for genes A, B and C, the bottom map is derived from a DNA study including a series of biochemical experiments (cutting the DNA, separating the fragments, hybridizing with the probes, etc.). The implicit assumption that the morphological characters A and B are each determined by a single gene is an oversimplification in this case.

The gradual refinement of genetic maps

One of the primary aims of the genome programmes is to define, or rather to refine, the genetic map. Notwithstanding the powerful techniques brought to bear, this remains a long and exacting task requiring numerous experiments and the use of advanced data processing resources. Headway on the genetic map is therefore slow. In 1987, a large group coordinated by Helen Donis-Keller [4] published a "ten centiMorgan" genetic map (i.e. a map in which landmarks are separated on average by ten centiMorgans) (Figure 2-2), and the announced target was for a one centiMorgan genetic map by 1992 – an ambition which proved far too optimistic and had to be scaled down. However, as we will discuss below, the worst is now over and work on the one centiMorgan map has been intensified, both because of an increased awareness of its importance (including as a reference for the physical map) and because of recent technological advances. These include the automation of Southern blots through the development of robots, each capable of performing several tens of Southern blots per day, and the widespread and intensive use of microsatellites [13] – simple sequences such as $(CA)n$ that are frequently encountered in the genome and extremely polymorphic. It thus appears that the rate of progress has picked up again for the genetic map after a slack period favouring the physical map.

This is a heartening trend as a detailed genetic map is absolutely essential for medical genetics. For instance, take the case of a disease whose hereditary nature is discovered (by the clustering of successive cases in certain families) but about which nothing else is known. Then the only way – albeit very time-consuming – of localizing the gene responsible is to study the relationship between the disease and the landmarks on the genetic map: this requires a pre-existing map, the more detailed the better ! Once this is achieved and though the incriminated gene must still be isolated and identified, at least its approximate whereabouts are known.

Physical maps make headway

The second major objective of the genome programmes is detailed physical mapping of whole chromosomes. This is achieved by constructing maps using pulsed-field electrophoresis, by cloning large DNA segments as yeast artificial chromosomes (YAC), and by lining up hundreds or thousands of cosmids or YACs in sets of overlapping, contiguous clones (often called "contigs"), following the path traced out by Y. Kohara several years ago for the *Escherichia coli* genome ("only" five million bases) (Figure 2-3). The necessity of investing in this kind of work is self-

EMERGING HUMAN GENOME PROGRAMMES 9

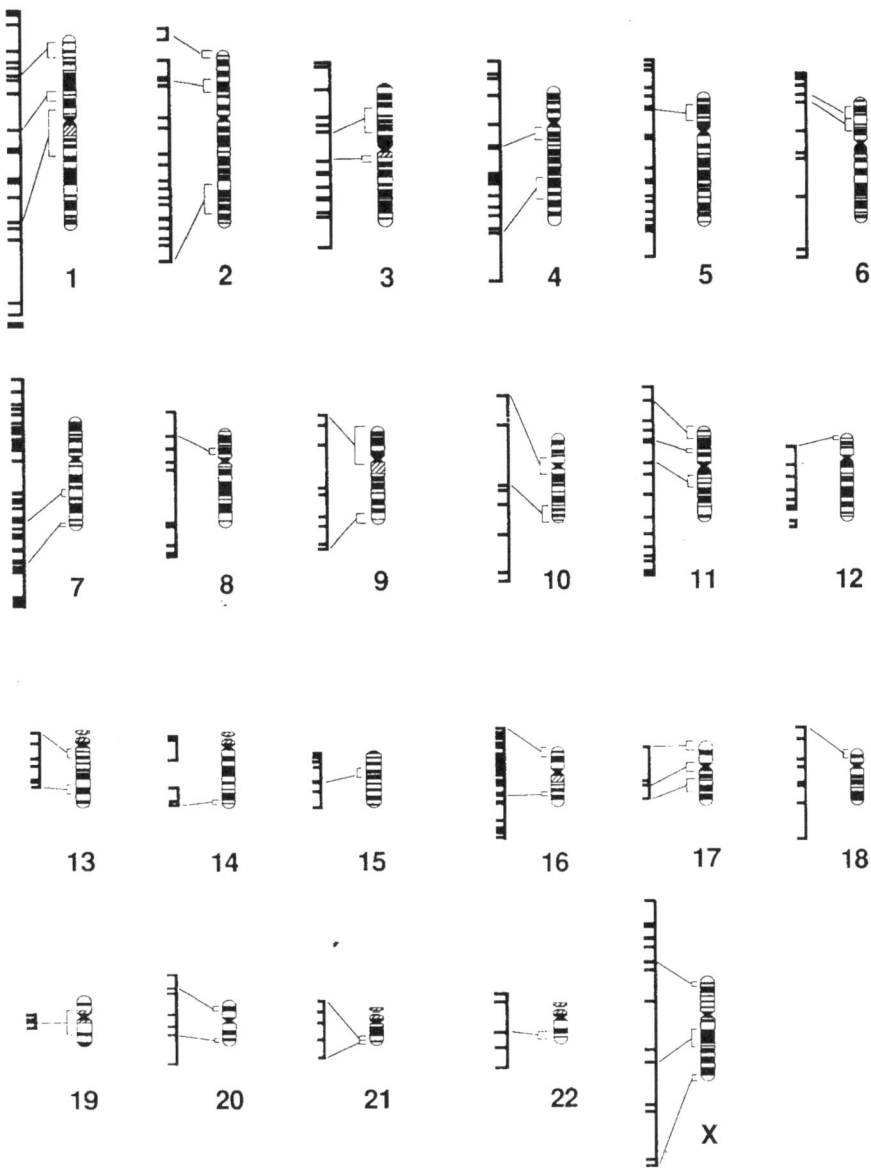

Figure 2-2 The first human genetic map. Published in October 1987 by Helen Donis-Keller and many coworkers, this map appeared on the prestigious cover page of *Cell*. It nevertheless drew criticism from many scientists on the grounds that its publication was premature in that the maps of some chromosomes were very incomplete.

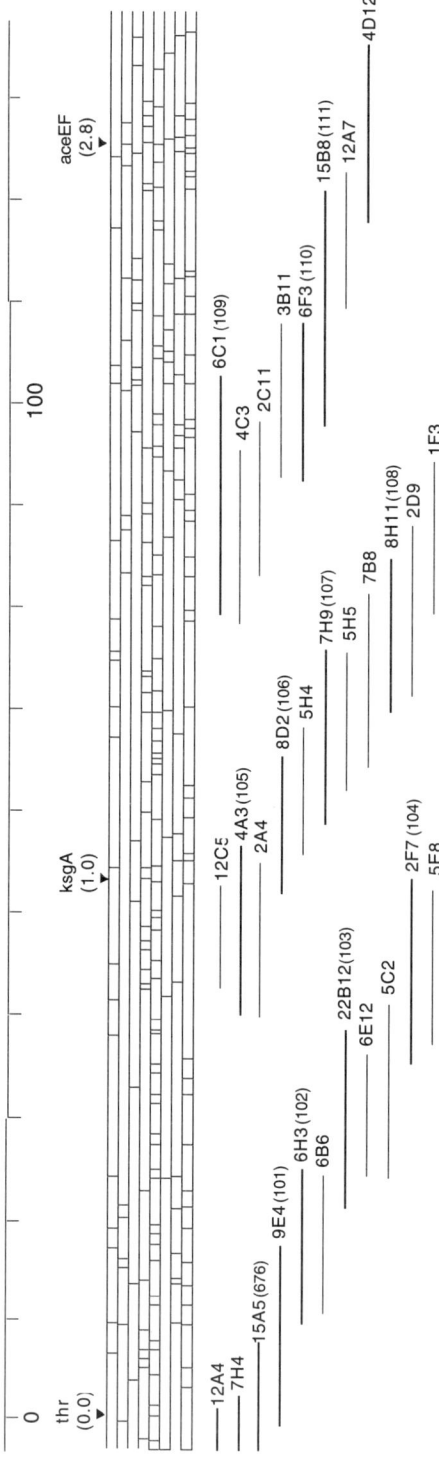

Figure 2-3 A small part of the physical genome map of the bacterium *Escherichia coli*. This map was constructed by Kohara and his collaborators in 1987. The illustration shows about 1/40th of the DNA molecule. The line at the top is the scale in kilobases and the central part (like a stave in music) indicates the sites cut by eight common restriction enzymes. The staggered strokes below indicate the position of each of the cloned DNA segments (in the lambda bacteriophage) used to construct the map. The letters above the "stave" specify the positions of previously known genes.

evident. Prior to the extremely detailed examination provided by DNA sequencing (the ultimate form of the physical map), the area must be marked out and landmarks positioned for a complete but less detailed survey. Premature large-scale sequencing would prove of little use and would generate a jumble of sequences without enough information on their precise location on the genome. Physical mapping is thus a structuring phase essential to the next step, sequencing which, if desired, can then be carried out selectively. As a rough estimate, more than half the funds of genome programmes are devoted to this type of work. Very significant progress has been achieved recently, thanks in particular to creative approaches based on the use of YACs, and the first whole-chromosome maps were published at fall 1992 [1, 8].

Sequencing: methodology and cDNA

DNA sequencing is indeed an important part of present-day Genome programmes, but in terms of technological improvement as much as of data collection. This technology factor is in fact a common denominator throughout the programme and it is easily transferable to other sectors. Improvements to sequencing have been sought in several directions: by direct automation of existing manual techniques (as practiced in Japan) or, in a more innovative spirit, by developing new technologies. Some of these are relatively conventional, such as the multiplex sequencing method of George Church [2] aiming to extract much more information from a sequencing gel, or capillary electrophoresis that should extend the number of nucleotides read in a single run; others are as exotic as tunnelling microscopy, which could allow the bases of a DNA molecule to be directly "viewed", thereby making it possible to read the sequence directly [10]. At the same time, computer programmers (or computer-literate biologists) are perfecting their software to interpret the sequence data and to recognize genes in them. This problem is only partially solved because the signals characterizing genes are often fuzzy, leading some specialists to think that this need could only be fulfilled by neural networks; in any case, this is a very important field, receiving a lot of attention.

A recent trend emphasized in the British, Japanese and French projects – but in fact undertaken most efficiently in the USA – has led to partial but massive sequencing of complementary DNA clones, thereby placing priority on expressed sequences. This line of research has proved very promising because of its efficiency. Many laboratories are therefore working on this topic, giving rise to much lively debate on the possibility and desirability of patenting these sequences, or more precisely their potential applications.

National policies:

USA...

Any discussion on Genome programmes is hardly possible without talking about the "heavyweights" in the field (DOE, HHMI, HUGO, etc.) and of course giving a plethora of figures, preferably in millions of dollars. We will first briefly review the current programmes and the money and organizations involved. The specific situation in each country will be discussed later in detail, but the background information given below will be useful to follow the next chapters.

In the USA, leadership of the project was taken up in 1986 by the Department of Energy (DOE), an agency which has responsibility for both civilian and military nuclear programmes. Though the link with the human genome is not obvious, DOE has in fact been involved for a long time in biological research, notably on cellular systems which repair radiation damage in DNA. DOE, which has expertise in high power lasers (developed for Star Wars applications), also put this to use for sorting chromosomes, constructing specific DNA libraries and providing them to the scientific community: these are the so-called Livermore and Los Alamos libraries, today distributed by the American Type Culture Collection (ATCC). In short, DOE has definite high tech biological know-how; it has heavily invested in the Human Genome Programme whose technological and methodical characteristics appear particularly suited to its very structured organization.

Although somewhat reticent at the outset, the National Institutes of Health (NIH) quickly claimed a role in the undertaking. NIH is a Federal agency with a budget of $9,000 million (20 times that of the major corresponding French agency, INSERM) that awards grants but also runs "in-house" laboratories (e.g. at Bethesda near Washington). NIH specializes more in fundamental biology than DOE and it was quick off the mark to claim a major role in the programme. Last of all a third participant in the genome research programme, particularly in the starting phase, was the Howard Hugues Medical Institute (HHMI), founded with the fortune of the eccentric billionaire of the same name.

In fact a working arrangement has been found according to which NIH provides roughly $100 million to fund genome investigations – in preference directed towards fundamental and academic aspects – as part of a programme directed by James D. Watson up to mid-1992. As for DOE, it devotes an almost similar sum to the same topics but with more emphasis on technology (e.g. robotics, new sequencing methods, etc.) in its three centres at Los Alamos, Lawrence Livermore and Lawrence Berkeley. Last of all, HHMI funnels roughly $30 million to a few selected laboratories throughout the USA. The total sum invested is thus substantial, amounting to more than $200 million (i.e. half the budget of INSERM including salaries). Moreover, this funding is a long-term effort planned over 15 years or so. Thus the current supremacy of the United States in this sector, as highlighted in Figure 2-4, does not come as a surprise.

Japan...

The situation in Japan is different. Several somewhat futuristic publications, the announcement of the Human Frontier Programme (commonly – and erroneously – believed to deal with research on the human genome), and the reputation of Japanese technology led many to believe that this country was playing a very significant role in the Human Genome Programme in the nineteen-eighties. In reality this was not true and, in terms of genetics, Japan is still in fourth place behind France (Figure 2-4). This position is now evolving and Japan may become a leading player in the Human Genome Programme.

Europe...

What about the Old World ? Here Great Britain occupies a place apart due to its strong tradition in molecular biology and human genetics combined with state of the art technology. A specific programme was initiated in 1989 under the impetus of Walter Bodmer, Sydney Brenner and others. The funding, £11 million (more than $20 million) over three years cannot in any way be termed insignificant. Furthermore, strong research foci already exist (Imperial Cancer Research Fund (ICRF), Cambridge, Oxford, Edinburgh, etc.) and a service centre has been set up to distribute probes, libraries and cells. Another positive factor is the special relationship existing between the British and their transatlantic relations in the USA. In short, Great Britain is off to a rather good start in this race.

What can be said about the remainder of Europe ? Italy [5] has channelled its efforts on the distal half of the long arm of the X chromosome (Xq24-Xq28) and the programme seems quite promising. Germany and Sweden, among others, have now started specific human genome programmes, and this effort may be scaled up in the near future ; the Netherlands, Finland and others are also doing some excellent work. Up to recently France had no bona fide human genome programme and several first rate teams were performing relatively small-scale research in the field without major resources. CEPH (Centre for the Study of Human Polymorphism) was the only organization able to start up large-scale activities and to invest in technology. Founded in 1983 by Jean Dausset with the objective of forming a collection of families and distributing their DNA, CEPH plays a fundamental role in the construction of the human genetic map [3] and has implemented the necessary technologies for systematic research on the

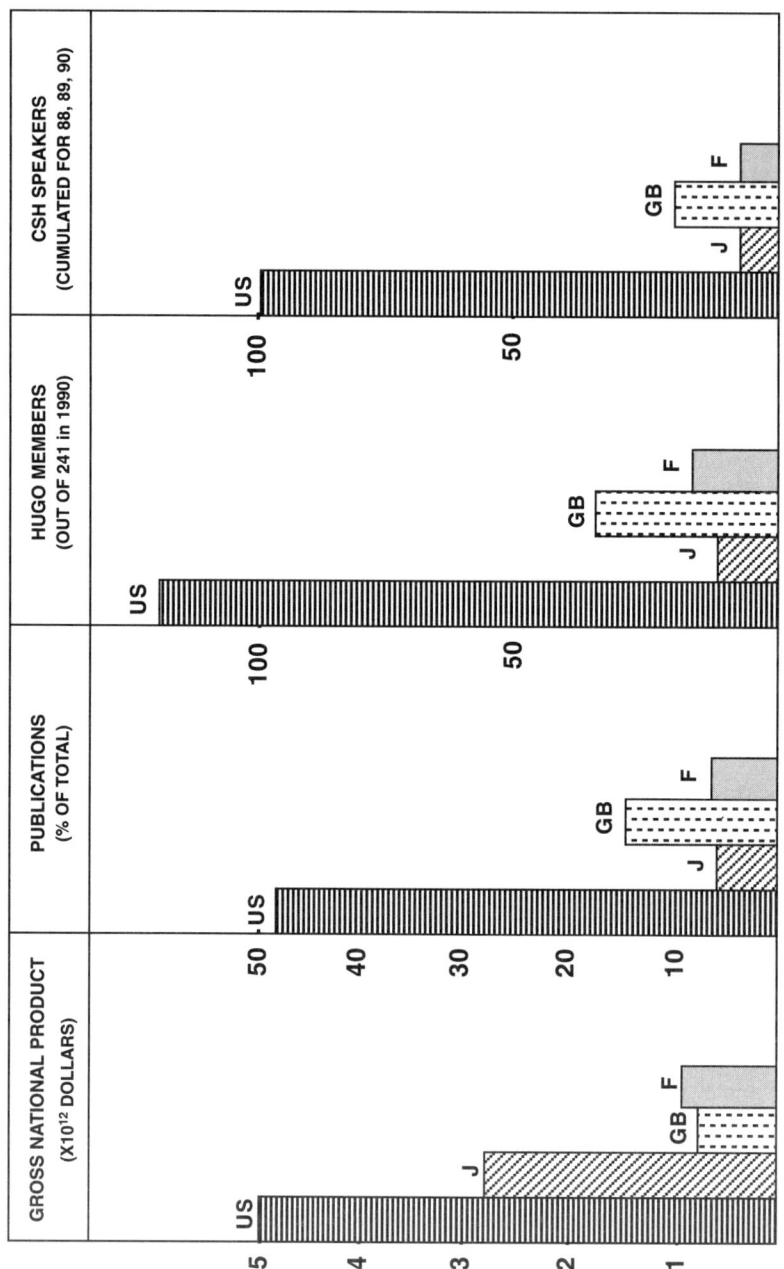

Figure 2.4 The genome "heavyweights". This figure presents four types of data for the four major countries contributing to genome research. From left to right: gross national product in millions of dollars; scientific output based on the share of articles on the human genome (this study was undertaken by the European Science Foundation); number of HUGO members as of March 1990 ; and last, cumulated number of speakers in the plenary session of the Annual Symposium on Genome Mapping and Sequencing, held at Cold Spring Harbor. A certain bias against non-Americans is evident in this last case...

genome. AFM (French Muscular Dystrophy Association) has organized several successful telethons and has allocated large sums to research on genetic diseases and, by extension, to the genome. AFM has set up an organization – the only one of its kind in the world – called the Généthon, designed to help laboratories work more efficiently by providing a pool of advanced equipment and technicians as well as to tackle some specific heavyweight genome projects. The latter have recently scored some spectacular successes, to be described later. To sum up, France possesses numerous structures capable of advanced research on the genome; their coordination, however, could definitely be improved.

The EEC...

In Europe, the EEC began preparing a research programme on the genome in 1987. First proposed under the rather unfortunate name of "Predictive Medicine", this ran into strong criticism, especially from the German "Greens" who oppose genetic engineering and anything that could possibly be related to eugenics. This relatively modest programme (17 million Ecus, roughly 25 million US dollars) finally got underway in 1990 [7]. A second phase with a two-fold increase in funds started in 1993, emphasizing services (clone and data banks, resources for genetic mapping) rather than academic research, as is legitimate for a transnational project of this type.

... and HUGO

Last of all, genome research is also represented internationally by the Human Genome Organization (HUGO), founded in the spring of 1988 during the first Cold Spring Harbor meeting on genome mapping and sequencing. The aim of HUGO is to act as an international driving and coordinating force for this field. As such, HUGO should promote meetings and workshops, advise governments, and perhaps set up a system of fellowships, etc. A quick glance at the list of the first two hundred members (appointed by an international founding board) highlights the forces in action, i.e. over 120 members from the USA, more than 30 from Great Britain, about 15 from France; for most of the other countries members can be counted on the fingers of one hand. Though HUGO got off to a difficult start and took time to become effective in the world of the human genome, it nonetheless appears to be absolutely vital.

One point is clear: the United States and to a lesser extent Great Britain largely dominate this field (although recent developments in France may appear to threaten their supremacy). Though this is common to all biomedical research, it is particularly blatant in the case of the human genome. Apart from the programme's origin in the USA (not really a pure coincidence !) and the size (financial, etc.) of the USA, this disequilibrium stems from several factors. First, the organizational flexibility of research in North America allows resources to be rapidly mobilized for work on a given subject (the lack of such flexibility in French public laboratories is sorely felt). The lack of inhibitions with respect to industry (and industry's involvement in Research), the easy acceptance of technology, the extremely open attitude on the circulation of individuals and ideas, as well as the readiness to reassess leaders of organizations and even organizations themselves are all factors explaining the dominant position of the New World in this field.

REFERENCES

1. Chumakov I, Rigault P, Guillou S, Ougen P, Billaut A et al: A continuum of overlapping clones spanning the entire human chromosome 21q. *Nature* 1992, **359**: 380-387
2. Church GM, Kieffer-Higgins S: Multiplex DNA sequencing. *Science* 1988 **24**: 185-188
3. Dausset J, Cann H, Cohen D, Lathrop M, Lalouel JM, White RL: Centre d'Étude du Polymorphisme Humain (CEPH). Collaborative genetic mapping of the Human Genome. *Genomics* 1990 **6**: 575-577
4. Donis-Keller H, Green P, Helms C, Cartinhour S, Weiffenbach B et al: A genetic linkage map of the Human Genome. *Cell* 1987 **51**: 319-337
5. Dulbecco R: The Italian genome program. *Genomics* 1990 **7**: 294-297
6. Ezzel C, Swinbanks D: Plant researchers eager for Genome programme. *Nature* 1989 **340**: 491
7. Ferguson-Smith MA: European approach to the human gene project. *FASEB J* 1991 **5**: 61-65
8. Foote S, Vollrath D, Hilton A, Page DC: The human Y chromosome. Overlapping DNA clones spanning the euchromatic region. *Science* 1992; **258**: 50-66
9. Jordan B: Au rendez-vous des cartographes. *Médecine/Sciences* 1989 **5**: 500-503
10. Jordan B: Le tunnel séquencera-t-il le génome? *Médecine/Sciences* 1990 **6**: 1007-1009
11. OTA report: Mapping our genes. The Genome projects: how big, how fast? BA 373, US Government Printing Office, Washington, 1988
12. Sinsheimer RL: The Santa Cruz Workshop. May 1985. *Genomics* 1989 **6**: 954-956
13. Weber JL, May PE: Abundant class of human DNA polymorphisms which can be typed using the polymerase chain reaction. *Am J Hum Genet* 1989 **44**: 388-396
14. White R: Chromosome Mapping with DNA Markers. *Scientific American* (1988) **258**: 40-48

3

Families, RFLPs and microsatellites: the tools for genetic mapping

Let us now return to the human genetic map and discuss the technical, logistic and organizational problems encountered in its development [1, 9], together with the prospects existing today. The genetic map is an inseparable part of the genome project: it represents the first survey providing a framework and starting point for the more accurate but much more labour-intensive studies of detailed physical mapping and their final outcome – the DNA sequence.

A map that needs refining

In 1987 a team coordinated by Helen Donis-Keller [3] – at the time with the US firm "Collaborative Research" – published the first general map of the human genome. The landmarks on this initial map had an average spacing of roughly ten centiMorgans (approximately 10% recombination, the total length of the human genetic map being close to 4,000 centiMorgans) (Figure 2-2, p. 9). The landmarks used are DNA probes, each corresponding to a point of the genome featuring sequence polymorphism, which causes restriction sites to appear or disappear when DNA is cleaved with certain restriction enzymes. Thus the probe will detect different fragments in different individuals – hence the term RFLP or Restriction Fragment Length Polymorphism (Figure 3-1).

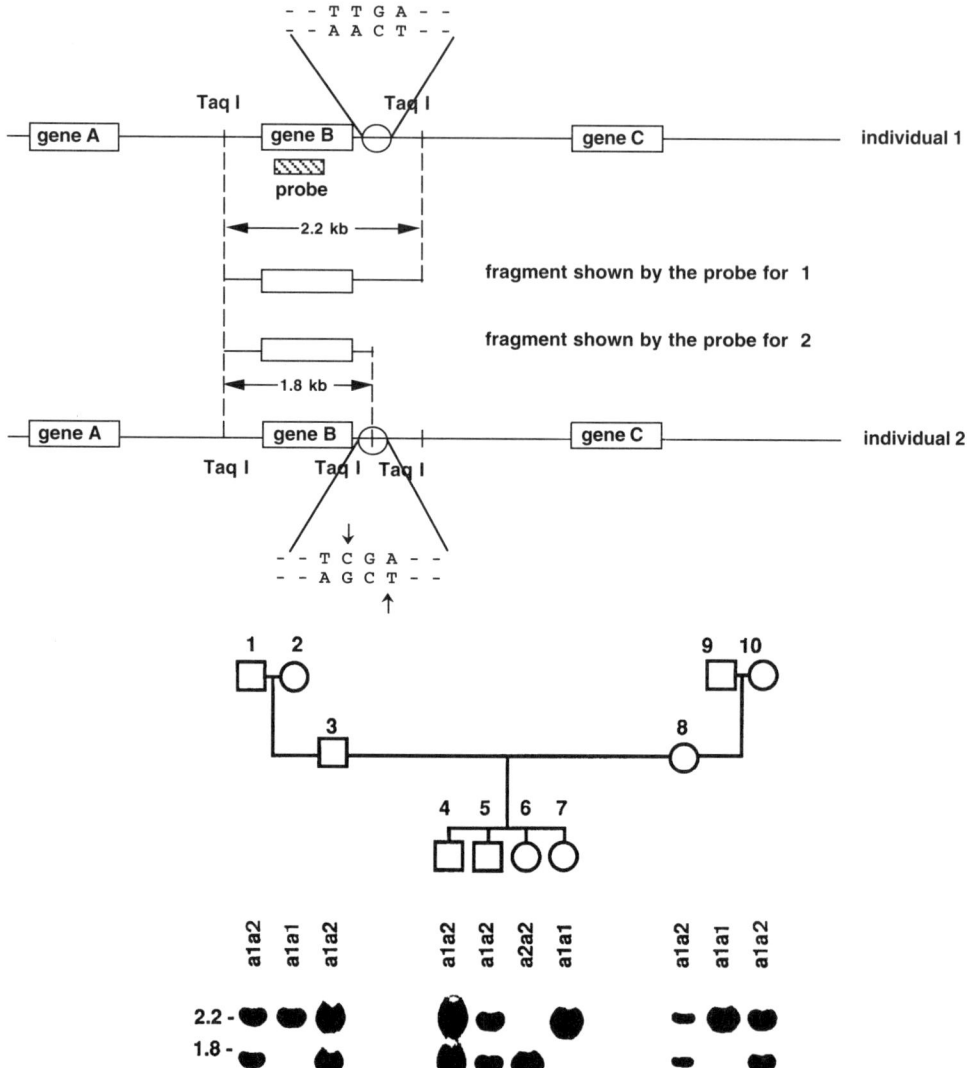

Figure 3-1 Restriction polymorphisms and their utilization. Detection of a DNA polymorphism resulting in size variation of the fragment detected by a probe (RFLP for Restriction Fragment Length Polymorphism). A nucleotide substitution causes a site for the Taq 1 restriction enzyme to appear or disappear. Due to the location of the probe (hatched) used in the Southern blot, this polymorphism will result in a fragment of 2.2 kilobases (2nd Taq I site absent, at top) or of 1.8 kilobase (2nd Taq I site present, at bottom). The lower half of the figure shows the result of an analysis in a family in which some individuals are heterozygous (alleles 1, 2.2 kilobases and 2, 1.8 kilobase; the probe is assumed to detect a locus on an autosome). DNA analysis (cleavage by Taq I enzyme, migration on agarose gel, transfer to a membrane and hybridization with the probe) on DNA from members of the family produces a series of bands, from which the alleles inherited by each of the children can be derived. When conducted on the same families with several probes corresponding to different loci, this type of analysis is the basis for construction of a genetic map.

Consequently, genetic mapping of our whole genome entails isolating a sufficient number of polymorphic probes – at least several hundred – and then studying how the respective alleles are transmitted from one generation to another. Thus with numerous probes and large, very well characterized families (hence the paramount role of the Centre for the Study of Human Polymorphism which has set up and made available to the scientific community such a collection of families), it will be possible to determine the relative positions of the loci detected by the probes and to evaluate their distances as recombination frequencies (a centiMorgan corresponds to a recombination frequency of 1%).

Although the average ten centiMorgan spacing obtained by Helen Donis-Keller's team was a major step forward, the resolution was still much too coarse. One case in point is the use of "reverse genetics" (or "positional cloning") to search for the gene causing a disease. Thanks precisely to the genetic map and the above collection of probes, a genetic investigation of families with a given complaint allows one or two of the probes to be associated and linked with the disease. It is then possible to assert that the gene responsible (when defective) for the disease must be located somewhere between loci A and B defined by the corresponding polymorphic probes. All that remains to be done is to clone and analyze the DNA contained in that interval, to find the genes that are present there and to test each of them in patients and controls until the sought-after one is identified. This is the process successfully applied to isolate the genes involved in cystic fibrosis, neurofibromatosis and polyposis coli, among others. However, the problem is not as simple as it seems, since on average a genetic distance of one centiMorgan corresponds to 1,000 kilobases along the DNA, i.e. a million nucleotides. Thus if the two landmarks in question, A and B, are separated by ten centiMorgans, a 10,000 kilobases DNA interval will have to be cloned, dissected and analyzed – a task that still remains impossible or at best extremely time-consuming, even using the very latest methods of DNA analysis [6].

Disappointingly slow progress

It was thus not surprising to see priority given to producing a "one centiMorgan genetic map", in which the average spacing between landmarks would be one centiMorgan, corresponding roughly to 1,000 kb. In the USA, the National Research Council (NRC) allotted five years for the construction of this map. When appointed to lead the National Institutes of Health Genome Project in 1988, James D. Watson re-endorsed both the aim and the time frame. Notwithstanding, it rapidly became apparent that the project had set its sights too high and that progress was much slower than expected. This is quite

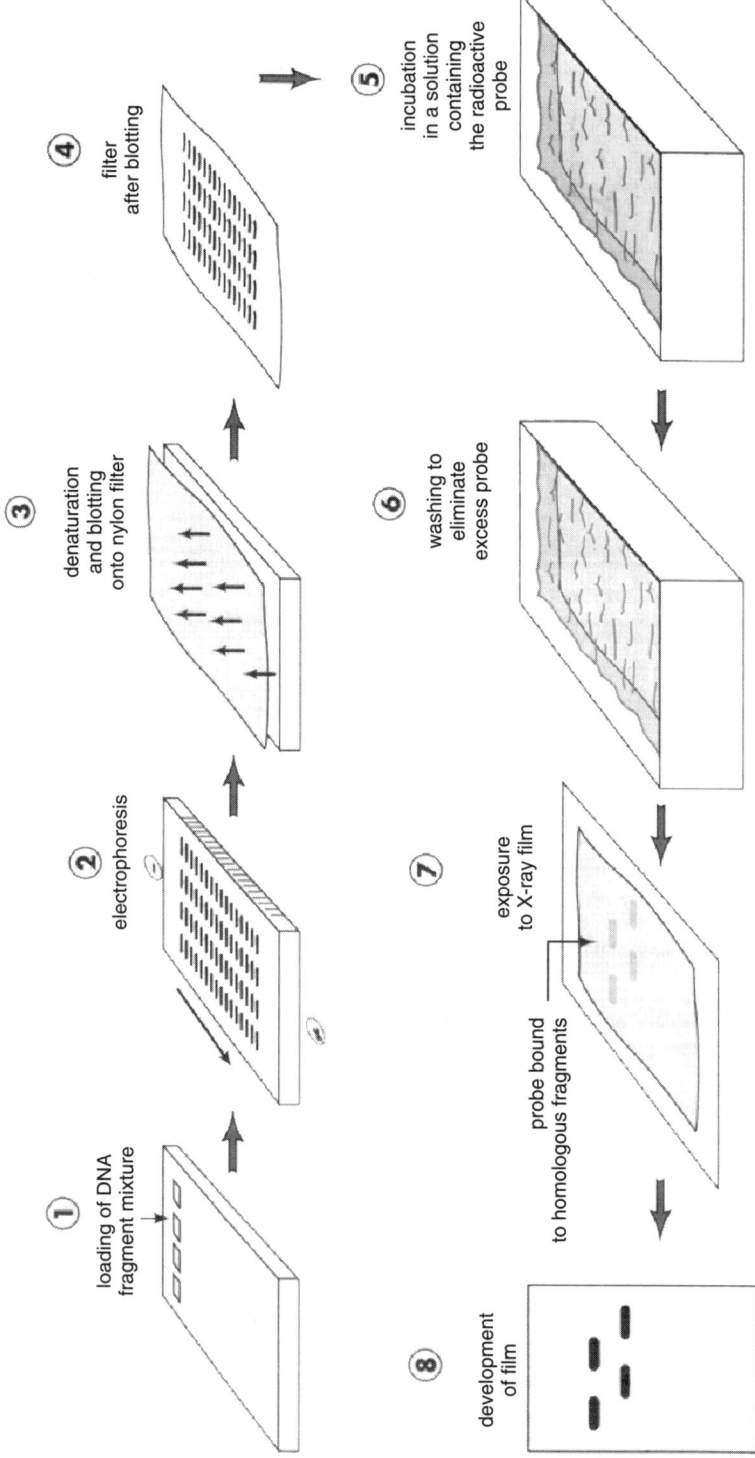

Figure 3-2 Schematic steps in Southern blotting. (Reproduced from "La génétique inverse", B. Jordan, Actuel Quillet, Paris 1991.)

Southern and Northern blots

The diagram on the opposite page illustrates the main steps in the Southern blot technique, named after its inventor Edwin Southern (Figure 3-2).

Step 1) The mixture of DNA fragments to be analyzed is placed in the well of an agarose gel. Usually this mixture is obtained by cleaving a total human DNA sample with a restriction enzyme. The mixture contains several hundred thousand different fragments whose size can run from several hundred up to several thousand bases.

Step 2) Separation by electrophoresis for several hours. The fragments migrate according to their size (the smaller they are, the quicker). The fragments may be visualized at this time using ethidium bromide staining but the complex makeup of the mixture results in a smear, rather than a series of bands as indicated in this schematic diagram.

Step 3) The gel is soaked in an alkaline solution which denatures the DNA. The two complementary strands separate *in situ* but do not move in the gel. A special filter made of nitrocellulose or activated nylon is placed in contact with the gel and an absorbent pad added on top. The liquid in the gel is drawn up by capillary action, carrying along the DNA fragments which are then trapped on the filter.

Step 4) The result is a print, a blot, of the gel. The previously separated DNA fragments are now fixed on the filter in a position matching their prior location on the gel and thus indicating their size. The fragments are denatured and therefore can bind to a DNA probe that contains the complementary sequence.

Step 5) The filter is now immersed in a solution containing the radioactive probe – a DNA segment previously tagged by incorporating radioactive phosphorus and denatured so that it is ready to bind onto a complementary sequence.

Step 6) After incubation at approximately 65°C during several hours to allow the probe to bind, the filter is washed to eliminate excess probe. After washing, the filter still contains the DNA fragments transferred in Step 3 but the radioactive probe has now bound to the filter at points with DNA fragments containing a sequence complementary to the probe (in fact at the start the separated DNA and the probe DNA are both double-stranded; they are then denatured, thus the probe "+" stand binds onto the fragment "-" strand and vice versa, the overall position on the filter remaining unchanged).

Step 7) When the filter is dry, it is placed in a darkroom in contact with a very sensitive photographic film – the type used for medical X-rays. The radioactivity captured at certain spots on the filter (the dotted marks in the diagram) will produce an image on the film.

Step 8) The developed film exhibits black bands corresponding to the places where the probe found complementary (and hence homologous) DNA fragments in the sample. Control fragments of known size integrated at the beginning in the gel allow it to be calibrated and the probe-detected fragments to be approximately sized.

In widespread use in molecular biology laboratories, the technique owes its success to its simplicity, its sensitivity (a single fragment can be detected out of the hundreds of thousands resulting from cleavage of total human DNA with a restriction enzyme such as EcoR1) and above all to its capability to search for such fragments simultaneously on a series of samples (usually 20 on each gel) and also to provide data on fragment size because of prior electrophoresis-based separation.

The Northern blot – so named because it is a sort of reverse Southern blot – applies the same principle to analyze an RNA mixture, generally consisting of messenger RNAs extracted from a cell tissue or sample. These RNAs are separated on the gel according to their size, then transferred onto a filter for hybridizing and autoradiography. This process can thus establish whether the sample contains RNAs complementary to the probe used, hence whether the gene corresponding to the DNA probe is active in the relevant tissue or cell. The process also determines the size of these RNAs.

understandable: if the technology used remains unchanged, a one centiMorgan map requires approximately a hundred times as much work as a ten centiMorgan map since ten times as many individuals must be studied with ten times as many probes. The resulting task is overwhelming, it is also particularly unrewarding. It involves many repetitive steps, starting with the restriction enzyme digestion of hundreds of DNA samples isolated from the cell cultures or blood samples coming from the individuals in the families surveyed. Gel electrophoresis is then used to separate the DNA fragments, followed by transfer onto membranes. At this point, each probe must be labelled and hybridized individually to the membranes; finally autoradiography is performed, followed by examination, comparison and interpretation of the results.

It is hard to imagine a more mind-numbing activity, especially as it must be performed with hundreds of probes on hundreds of samples. Given the existing organization, the major part of this work is currently carried out, even in the USA, by graduate students or young post-docs who understandably do not pursue this type of work for very long. In addition, despite the repetitive and highly systematic nature of the experiments performed, great care must be exercised to guarantee reliable results. To pursue methodical and apparently routine mapping under these circumstances is therefore proving difficult, and it is highly tempting, for instance, to focus on a specific region known to contain a gene involved in a disease and to specialize in this region. Although this approach is undoubtedly commendable and sometimes produces noteworthy results, no headway is being made during this time on the general map – a situation reminiscent of cartographers responsible for preparing an overall map of a country who would survey solely those sites liable to contain oil deposits !

Motivation, recognition, funding

The above analogy is far from unjustified. It highlights the difficulties inherent in pursuing a methodical but ungratifying analysis in an environment marked by the individualistic temperament of researchers, the race to publish results (and their promotion in the media) and decisions strongly influenced by trends and even fads. It has been estimated that the amount of effort invested by the ten or twenty teams competing for the identification of the gene involved in cystic fibrosis could have resulted, if directed differently, in the complete map of one chromosome or a major part of a chromosome. This is precisely one justification for a world-wide and systematic project, such as the Human Genome Project. But how can teams be continuously motivated to carry out this work without letting it drop when they strike the first rich offshoot ? This question is paramount and it will become all the more relevant when tasks such as systematic sequencing of large DNA regions are undertaken.

The self-discipline of researchers is not the only scarce resource: funding for refinement of the genetic map has also, it seems, been hard to come by at certain times. Even in the USA, agencies such as NIH, which had made drafting a one centiMorgan genetic map one of their priorities, seemed little inclined to finance this undertaking. A case in point was the refusal by the NIH study sections of several grant proposals submitted by Helen Donis-Keller, on the grounds that they were "lacking in originality". As Helen Donis-Keller herself remarked, "I never said it was innovative. But it is important and doable" [9].

Polymorphism: a question of degree

One additional hurdle lies in the low informativity of probes, a problem known from the start but which proved very troublesome. To be really useful in genetic analysis, a probe must be more than just polymorphic: it should be *extremely* polymorphic. In fact, any randomly selected probe can ultimately be considered as polymorphic because study of a sufficient number of persons will always, in the long run, find one of them with a sequence dissimilarity changing the size of a restriction fragment. However a polymorphism in which one allele exists in 99.9% of the individuals and the other in 0.1% will be of little utility as the probe will virtually never be informative in the families studied. As a general rule, the RFLPs used to construct the genetic map have two or sometimes three alleles with, in the best cases, equivalent frequencies. Nonetheless it is not unusual for them to be non-informative in a given family. When the two parents are homozygous for allele "a1" of the locus A (even though this allele is only found in 40% of the general population), this RFLP provides no information in the family under study...

The discovery of "minisatellites" by Alec Jeffreys' team [5] provided the beginning of an answer to this problem. These sequences, so named for historical reasons (under certain centrifugation conditions, they concentrate close to the so-called satellite DNA bands), consist of a number of repeated nucleotide patterns, each 15 or 30 nucleotides long – which explains their alternative designation as Variable Number of Tandem Repeats or VNTR, mostly used in the USA. Their appeal stems from the variability from one person to another of the number of repeats within a given minisatellite at a specific position in the genome. Thus a probe for a particular minisatellite is very polymorphic and reveals a locus featuring a large number of alleles, that are very likely to be informative in virtually any situation. In addition, a "general" minisatellite probe, hybridizing with a large number of different minisatellites due to their sequence homologies, will produce a pattern of bands on a Southern blot, similar to a bar code and frequently employed for identification and in forensic cases – these are the so-called "genetic fingerprints".

Microsatellites, *alias* CA repeats

The real solution, however, has been the discovery of "microsatellites", thus named because of their kinship with minisatellites; they are also called CA (or GT) repeats [10, 11]. These sequences consist of repeats of a very simple pattern. The most commonly encountered in man is (CA)n, where n may range from 5 to about 50. These sequences are very attractive because they are highly polymorphic: if present at a given point in the genome, they will occur as (CA)17 in one person, (CA)18 or (CA)15 in another and so on (Figure 3-3). As these sequences are extremely widespread in our DNA – on average one about every 10 kilobases – they act as perfect landmarks for a very high resolution genetic map.

There is a darker side to the picture: the detection of microsatellite polymorphisms necessitates more sophisticated technology than a straightforward Southern blot. The initial step is to locate a microsatellite in the area where a marker is needed, e.g. by screening cosmids known to originate from the region with a (CA)10 oligonucleotide: finding one or several is almost certain given their frequency. A DNA fragment containing the microsatellite is thus identified and then sequenced to define the unique sequence DNA on either side. This information then serves as the basis for synthesizing PCR primers directed to this specific point of the genome. DNA analysis of a series of individuals will thus consist of PCR amplification from each sample using these primers, followed by analysis of the amplification product on a high resolution acrylamide gel (of the kind used for DNA sequencing), which must resolve the variations in length of two nucleotides over a total length of between 100 and 200 (according to the primers selected). Although this system is clearly more complicated than the conventional Southern blot, most of it can be automated and, moreover, its power fully justifies its use since these markers are nearly always informative and very frequent. Consequently, microsatellites are being increasingly employed in the construction of the genetic map [4]; several thousand such markers have been characterized by Jean Weissenbach's group at Généthon, nearly a thousand of which form the basis for a "second-generation" genetic map [12].

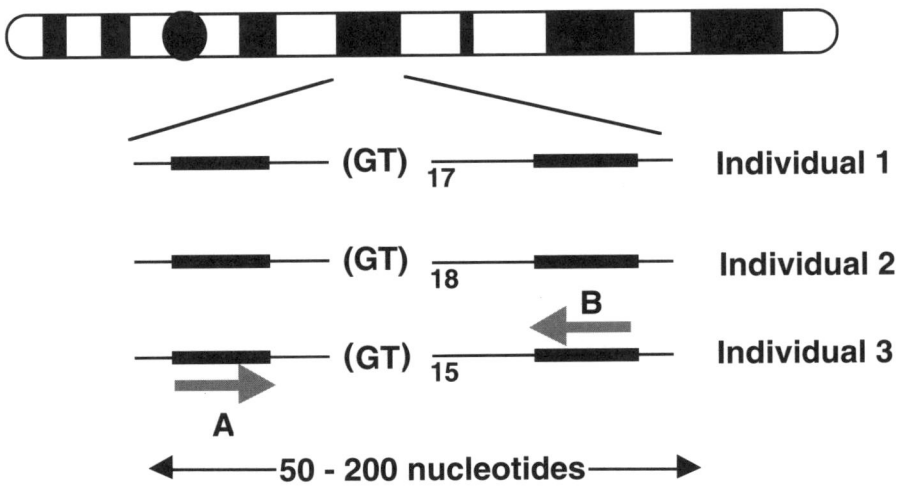

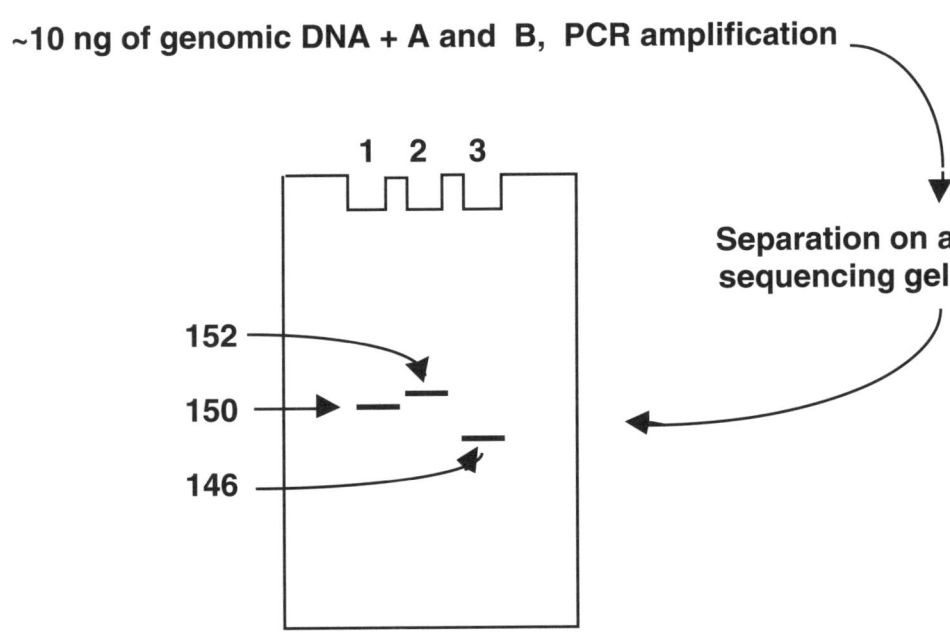

Figure 3-3 Microsatellites and their use. Analysis is based on PCR amplification using oligonucleotides A and B (grey arrows) that recognize single-copy sequences surrounding the microsatellites. After the reaction, the amplification products are analyzed on a high resolution acrylamide gel. With this gel it is possible to distinguish the fragment corresponding to the variant (GT)18 of the microsatellite from other fragments longer or shorter by a multiple of 2 nucleotides.

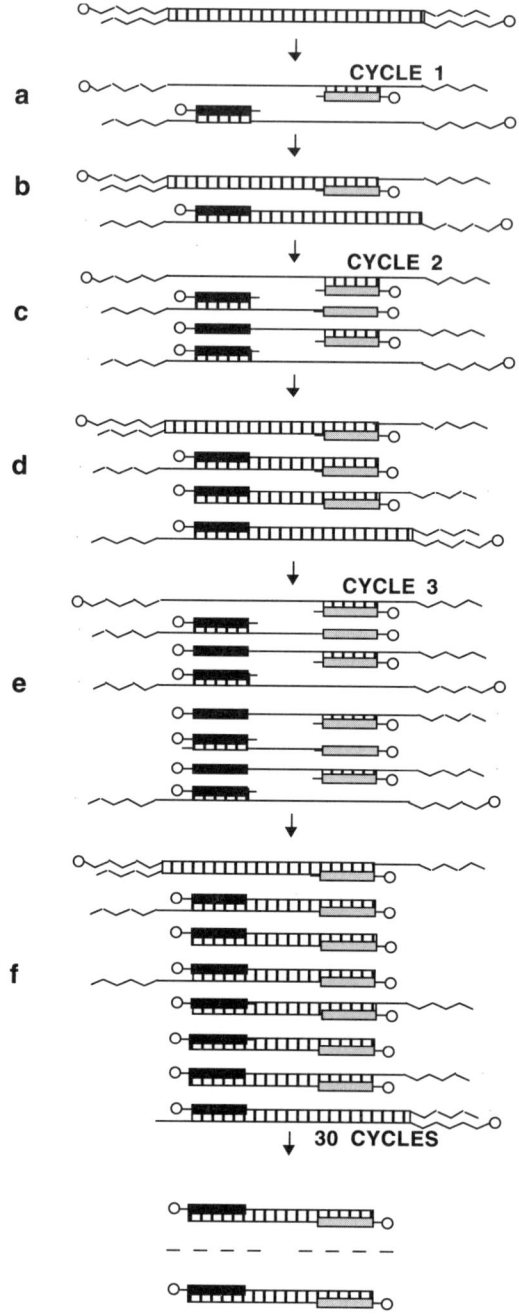

AMPLIFIED SEQUENCE ($10^5 - 10^6$ – fold yield)

Figure 3-4 Polymerase chain reaction (PCR)

PCR, Polymerase Chain Reaction

The polymerase chain reaction technique is probably the most important technical development that has occurred in Molecular Biology during the last few years. It makes possible specific amplification – up to a million-fold – of a small region of DNA within a very complex mixture; all that is needed is some sequence information on the segment to be amplified.

PCR relies on two oligonucleotides, complementary to each DNA strand and separated by a few hundreds or thousands of bases. These "primers" are synthesized, then annealed with the DNA sample in which they will specifically bind their complementary sequences (Fig. 3-4a). DNA polymerase is then added, it synthesizes the two complementary strands starting from the primers (Fig. 3-4b). A heating step denatures the DNA and separates the two strands, primers bind again on cooling, synthesis takes place again, yielding four new DNA strands, and so on (Fig. 3-4 c, d, e, f). Twenty such cycles provide a million-fold amplification.

In practical terms, amplification is performed with a thermo-stable DNA polymerase obtained from bacteria dwelling in geysers. Thus no enzyme addition is needed after the first step. The whole procedure is performed in thermocycling systems in which heating and cooling cycles can be programmed.

This technique, now ubiquitous, is extremely sensitive – hence its geat success. Well-chosen primers make it possible to amplify a specific human segment starting from total human DNA, a very complex sample (3 million kilobases). PCR thus supersedes cloning in a number of applications. To determine the sequence of the beta-globin gene present in a patient, for instance, it used to be necessary to construct a DNA library from DNA of this person, then to screen it to obtain the gene and finally sequence it. With PCR the relevant region can be directly amplified and sequenced starting from a few nanograms of total DNA (a drop of blood). Limitations of PCR include, of course, the need for some sequence information on the region to be amplified (although some tricks make it possible to by-pass this requirement in special cases); PCR is also limited – at least so far – to relatively short DNA segments, less than a few kilobases long. In addition, the extraordinary sensitivity of the method makes it very vulnerable to contaminations of samples, apparatus and reagents, and requires great care in laboratory practice.

Other innovations

Another new technique is based on the analysis of DNA contained in single sperm ("sperm typing"), though what impact it will have on the genetic map remains to be seen. It involves applying PCR to determine which allele of a given marker is present in the DNA of a single sperm isolated by micromanipulation (providing the local sequence is known, the PCR primers have been synthesized and the area defined by these primers exhibits informative polymorphism). This approach was first shown to be feasible in 1988 [7] and it has already been used to determine genetic distances [2]. Its advantages are obvious: the sperm is haploid, thereby considerably facilitating the analysis. In addition, each sperm contains chromosomes produced by recombinations independent of those occurring for the chromosomes of another sperm. Examining 100 sperm cells is thus equivalent to analyzing 100 children from the same father – a providential situation (at least for the geneticist), that rarely occurs in the real-life world ! This technology is particularly well adapted to the measurement of small genetic distances, at the price of some technical challenges (micromanipulation, amplification using very small amounts of DNA). Two or three loci – or perhaps more – may be simultaneously typed. It initially seemed unlikely, however, that a large number of loci could be studied in a given sperm. However the same authors have recently published a new method in which all or most of the sperm DNA is amplified using a set of random sequence primers [13]: nanogram quantities of haploid DNA are thus generated, and this removes the major limitation of the approach. It will be very interesting to watch the impact of this procedure on the general genetic map...

An additional hope for the genetic map lies in automation of very repetitive procedures like Southern blotting. Advances are being made in this automation, though many hurdles remain to be overcome. The most advanced implementation in this field is surely the Généthon at Evry where 20 machines can produce a total of 320 Southern blots at a time (Figure 3-5), and in fact have provided membranes for the European "EUROGEM" genetic mapping project. Thus it should be possible to boost the number of analyses, performed in good conditions of reliability – although costs remain a problem.

On the way to a two centiMorgan map

A glance at specialist journals reveals the progress accomplished, with improved versions of genetic maps for various chromosomes being published regularly, as a result of work in "consortiums" often using the families collected by CEPH. These maps, being more and more detailed (Figure 3-6), are consequently increasingly useful in the search for disease-generating genes. Two general genetic maps have recently been published, the first to appear since 1987 [3]. One of them, the NIH/CEPH map [8], results from the collaboration of many groups using the CEPH panel of families; the other one, derived by Jean Weissenbach's group at Généthon (France) [12], is a wholly CA-repeat-based map and as such both homogeneous and very useful because of the high polymorphism of these markers. Nevertheless, the genetic map alone is not sufficient and a reference system more directly anchored in the DNA is required: this is the aim of the physical map.

Figure 3-5 The Mark II Généthon room. This room is but one of the components of the Généthon, though surely the most spectacular with its 20 robots – in fact copies of the second Bertin/CEPH prototype developed within the framework of the European LABIMAP 2000 project. The top photograph shows several of these machines installed around the central equipment area containing the power supplies and cooling systems. In the lower photograph, some components of the Mark II machine are visible, i.e. the tank containing 16 vertically arranged gels, a manipulator arm for carrying out the enzyme digestions and loading the samples prior to separation by electrophoresis, and a microcomputer that controls these operations. On the right hand side, the cassettes into which the agarose gels are poured in direct contact with an activated nylon filter. With this set-up, crosswise electrodes are used to transfer the DNA, after electrophoresis, inside the tank without change of position or buffer. A commercial version of this machine has been marketed at the end of 1992.

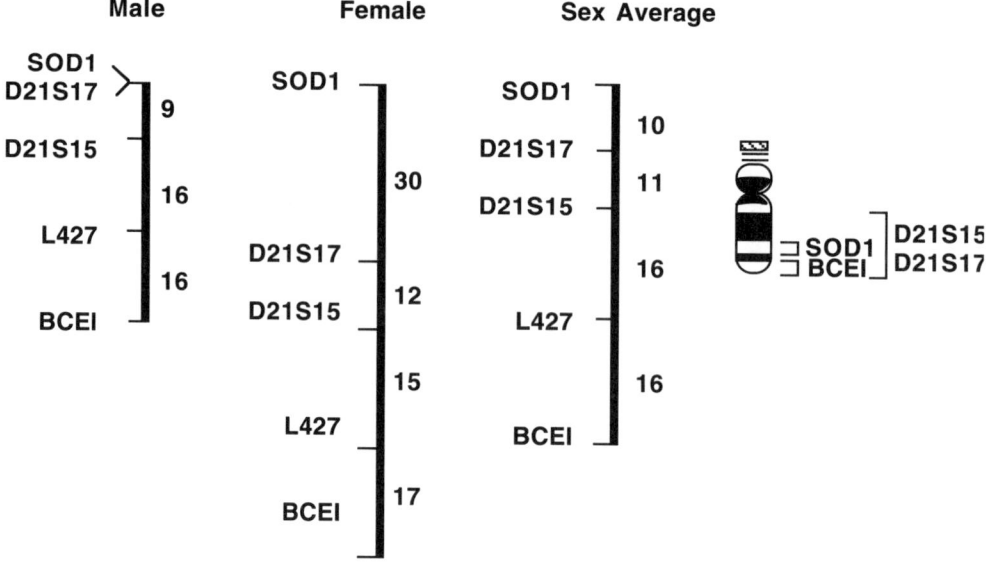

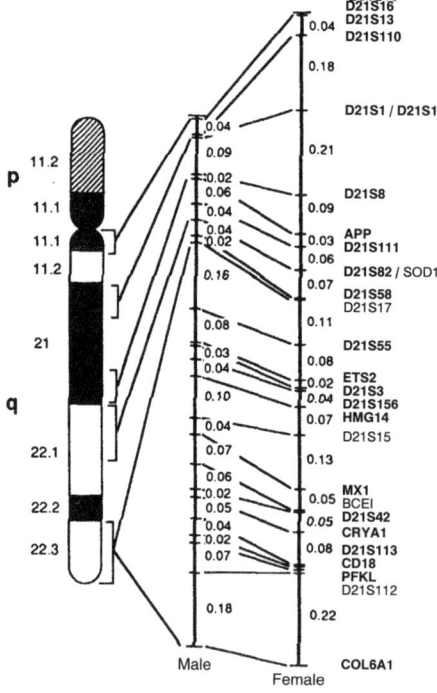

Figure 3-6 Genetic map of chromosome 21 in 1987 and 1991. The top half of the figure shows the genetic map of chromosome 21 in 1987, the lower half the same map in 1991. Both male and female maps are given since recombination frequencies are different even though physical distances are the same. The 1991 map is clearly much more complete. The SOD1 (superoxide dismutase) and BCE1 (translocation point) loci have moved from the end of the chromosome to inside, several telomeric loci having been found in each case. In contrast D21S15, an anonymous probe roughly in the middle of the chromosome has hardly moved at all.

REFERENCES

1. Anderson GC: Creation of linkage map falters, posing delay for genome project. *The Scientist* 1990 **4:** 1-13
2. Cui XF, Li HH, Goradia TM, Lange K, Kazazian HH Jr et al: Single sperm typing: determination of genetic distance between the G-globin and parathyroïd hormone loci by using the polymerase chain reaction and allele-specific oligomers. *Proc Natl Acad Sci USA* 1989 **86:** 9389-9393
3. Donis-Keller H, Green P, Helms C, Cartinhour S, Weiffenbach B et al: A genetic linkage map of the Human Genome. *Cell* 1987 **51:** 319-337
4. Hazan J, Dubay C, Pankowiak MP, Becuwe N, Weissenbach J: A genetic linkage map of human chromosome 20 composed entirely of microsatellite markers. *Genomics* 1992 **12:** 183-189
5. Jeffreys AJ, Wilson V, Thein SL: Hypervariable "minisatellite" regions human DNA. *Nature* 1985 **314:** 67-73
6. Jordan B: Megabase methods, a quantum jump in recombinant DNA techniques. *Bioessays* 1988 **8:** 140-145
7. Li HH, Gyllensten UB, Cui XF, Saiki RK, Erlich HA et al: Amplification and analysis of DNA sequences in single human sperm and diploïd cells. *Nature* 1988 **335:** 414-412
8. NIH/CEPH collaborative mapping group. A comprehensive genetic linkage map of the human genome. *Science* 1992 **258:** 67-80
9. Roberts L: Whatever happened to the Genetic map? *Science* 1990 **247:** 281-282
10. Tautz D: Hypervariability of simple sequences as a general source for polymorphic DNA markers. *Nucleic Acids Res* 1989 **17:** 6463-6471
11. Weber JL, May PE: Abundant class of human DNA polymorphisms which can be typed using the polymerase chain reaction. *Am J Hum Genet* 1989 **44:** 388-396
12. Weissenbach F, Gyapay G, Dib C, Vignal A, Morissette F et al: A second generation linkage map of the human genome based on highly informative microsatellite loci. *Nature* 1992 **359:** 794-802
13. Zhang L, Xiangfeng C, Schmitt K, Hubert R, Navidi W et al: Whole genome amplification from a single cell: implications for genetic analysis. *Proc Natl Acad Sci* USA 1992 **89:** 5847-5851

4
Evolving tools for physical mapping

To physically map the genome, or less ambitiously a chromosome or a chromosome region, entails – as with the genetic map – defining landmarks and locating them with respect to each other. The physical map is however much more directly related to the actual structure of the DNA molecule than its genetic counterpart. The landmarks used are directly embodied in the DNA, i.e. they are previously cloned genes or "anonymous segments", pieces of DNA that do not (apparently) correspond to genes. They need not reveal polymorphism, which is quite irrelevant in this context. A second feature of the physical map is that distances are also part and parcel of the DNA structure, the unit being the base (or nucleotide or the basepair which amounts to the same) and its multiples, the kilobase (1,000 bases) and even the megabase (1,000 kilobases or 1,000,000 bases) – rather than recombination fractions or centiMorgans. The construction of the map no longer involves studies on the transmission of alleles in families, but rather a series of biochemical operations, i.e. cleavage of DNA, analysis of the fragments, determination of their junction points. In other words, a sort of jigsaw puzzle whose completion will precisely fit together all the various parts along the DNA fibre.

Complementary methods: *in situ*, Southern blots and cloning

Several physical mapping methods can contribute to this task. *In situ* hybridization is special in that it is based on binding a probe by molecular

Figure 4-1 Two key scientists involved in physical mapping of DNA. At the top, Charles Cantor who together with Charles Schwartz invented the pulse-field gel electrophoresis technique. At the bottom, Maynard Olson (hands in pockets), jointly responsible with David Burke and Georges Carle for developing yeast artificial chromosomes. (Photographs taken by the author during a Genome Mapping and Sequencing Symposium at Cold Spring Harbor.)

hybridization, not to a gel or filter but to methaphase chromosomes (colour plate A, B). It can quickly determine the region of origin for a given probe but its accuracy is limited since the localization thus obtained is within one or half a band (equivalent to one or several megabases). The accuracy can be improved in various ways, notably by performing the experiment on interphase nuclei (colour plate C) though the chromosomes cannot be recognized in this case. This problem can be circumvented by simultaneous hybridization of several colour-coded probes to yield relatively accurate data, e.g. the order of DNA segments 100 kilobases apart. The usefulness of these methods has been tremendously enhanced in recent years through the replacement of radioactive labelling by non-isotopic methods, which have provided an increase of at least one order of magnitude in speed and precision while reducing considerably the labour involved in the procedure.

The other physical mapping methods make more direct use of DNA analysis techniques and of cloning. In essence, DNA analysis employs restriction enzymes to cleave the DNA into fragments that are then separated by electrophoresis, transferred to Southern blots and hybridized with probes from the region explored. Cloning, on the other hand, isolates one specific region (colour plate D) from the remainder of the genome so that it can be examined in detail (i.e. up to and including sequencing) and used as a probe. When these two complementary methods are combined, they can be used for chromosome walking, a procedure in which the cloned segments representing a given region are successively isolated in a stepwise process.

A problem of scale

While the theoretical course of action outlined above appears promising, a problem of scale is encountered in real life. The complete human genome (24 chromosomes end to end) spans about 3,000 million nucleotides, in fact close to two meters of DNA compacted into a cell nucleus whose diameter is roughly 10 microns. The X chromosome, for instance, contains a DNA molecule of 150 megabases and has a genetic length of approximately 180 centiMorgans. In other words, markers located one centiMorgan apart on a very detailed genetic map will therefore be spaced on average at one megabase intervals. Thus the physical map must be based on landmarks separated by one to several megabases, whereas until recently only DNA segments 20 or 30 kilobases long could be handled by the most advanced methods available: electrophoresis on agarose gel and cloning in phage or cosmid vectors. This major handicap has fortunately been largely eliminated in the last few years by the development of pulsed field gel electrophoresis (capable of analyzing DNA fragments from several hundred to several thousand kilobases) and, more recently, by the YAC technique whereby such large DNA segments can be "dressed up" as yeast chromosomes and then cloned in yeast [5].

Megabase methods:

Pulsed field gel electrophoresis for analysis...

Conventional electrophoresis in agarose gels is capable of separating small DNA segments – up to 20-30 kilobases long – according to their size; it is based on their differential rate of movement through the gel pores. This principle of separation no longer works when the DNA fragments are very large molecules measuring 100, 500 or 1,000 kilobases. The length of such fragments is much greater than the diameter of the pores in the gel; they move through the gel by threading through the pores in a "reptation" mode which gives them all the same speed. No separation is therefore achieved. This problem was overcome by the development of pulsed field gel electrophoresis by Schwartz and Cantor, in 1984 (Figure 4-1), [22]. Using an electrical field whose direction changes periodically (Figure 4-2), this technique forces the molecules to realign frequently. Large molecules take more time to do this than smaller ones: a very long DNA molecule (1,000 kilobases) will spend most of the time realigning, a smaller one (100 kilobases) will mostly be migrating and will thus have travelled much farther at the end of the electrophoresis run. This application of this technique allows the separation of DNA fragments of up 1,000, 2,000 or even 10,000 kilobases, making possible the construction of restriction maps (Figure 4-3) which, given the size of the segments analyzed, span thousands of kilobases and are thus suitably scaled for correlation with genetic linkage data.

... and YACs for cloning

A discovery that comes at the right time

In 1987 Burke, Carle and Olson published their seminal paper [5] reporting that large segments of human DNA can be propagated as Yeast Artificial Chromosomes (YACs) (Figure 4-4). As could be expected, these results dit not go unnoticed since cloning of DNA segments several hundred kilobases long was clearly a decisive breakthrough in the study of the human genome

[14]. The news was greeted with enthusiasm and many laboratories set out to construct YAC libraries using the vectors and strains generously provided by Burke and coworkers.

Initial results were not up to expectations, and it turned out that construction of YAC libraries is a rather tricky process. Yeast is not handled as easily as *Escherichia coli*, the preparation of very large DNA segments is not trivial, and the first libraries obtained by those who set out to transfer the method to their laboratories generally contained mostly clones with disappointingly small inserts, ranging from 50 up to a maximum of 150 kilobases. Nevertheless, progress was gradually made; today a handful of laboratories master this

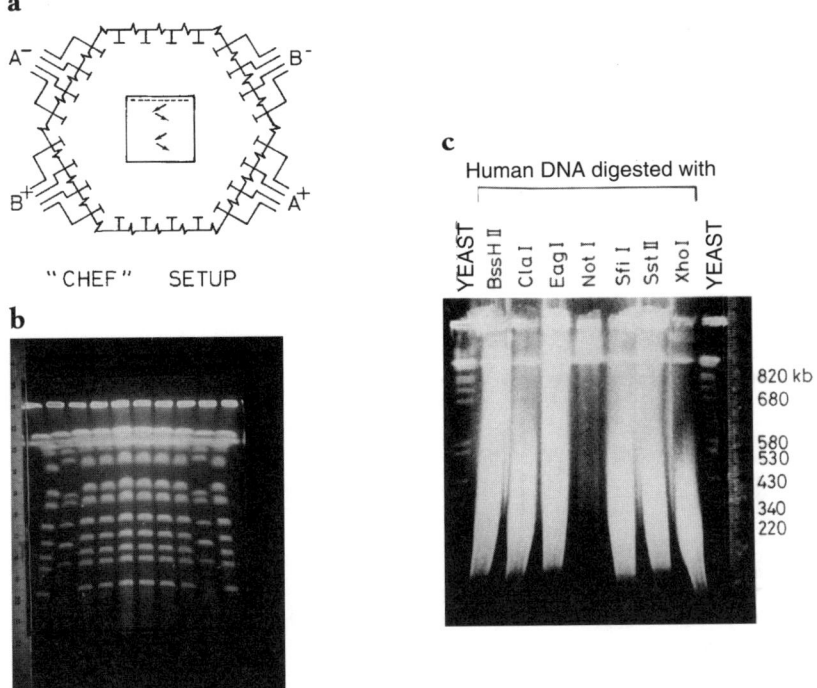

Figure 4-2 Separation of DNA fragments by pulse-field gel electrophoresis. One of the most commonly used set-ups – the CHEF arrangement – is shown at upper left. The electrodes are placed so as to alternatively generate uniform electric fields along the two diagonals. The gel is shown in place and the arrows indicate the directions of the successive migrations of the DNA molecules. The photograph at the lower left shows the separation obtained for yeast chromosomes varying in size from 100 to 2,000 kilobases. The lower right photograph illustrates an experiment in which human DNA cleaved by rare site restriction enzymes has been analyzed by this method, with the yeast chromosomes acting as migration controls. The individual fragments in the tracks containing the human DNA are not visible because this DNA is so complex that even such an enzyme produces several thousand different fragments. After separation the gel may be transferred by Southern blot and the resulting membrane hybridized with various probes to construct the DNA map in the vicinity of the probes (Figure 4-3).

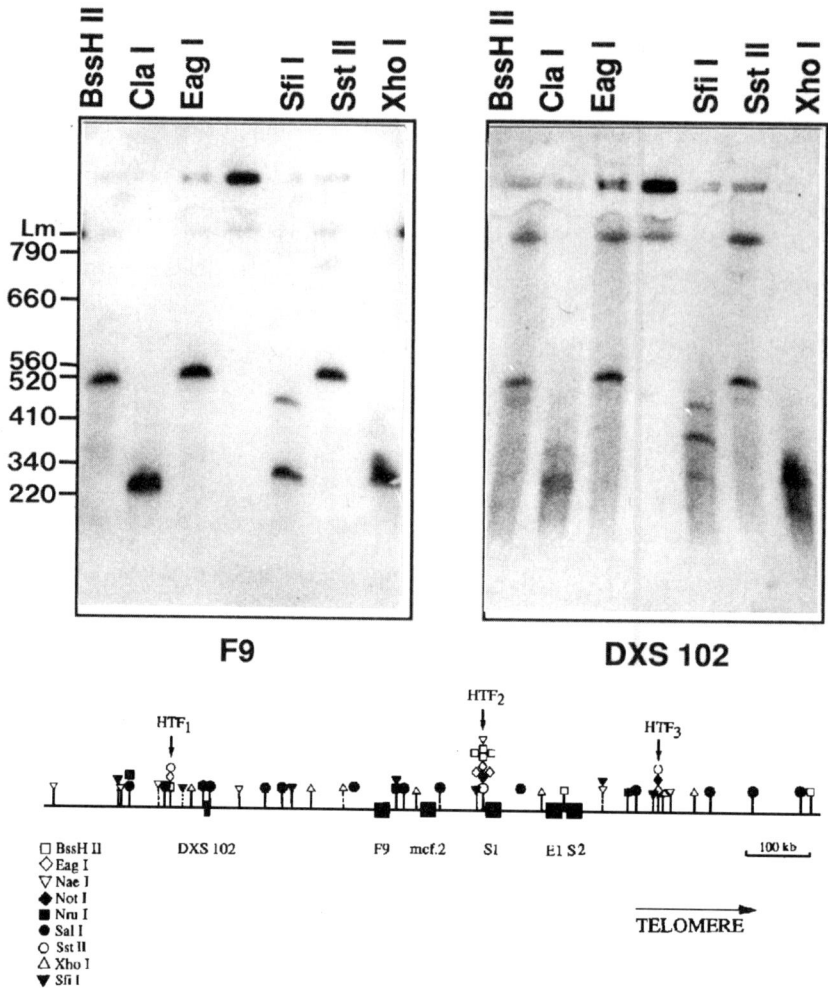

Figure 4-3 A typical physical map constructed on the basis of pulse-field analyses. The map shows a region of chromosome X spanning approximately 1,500 kilobases that was studied by pulse-field genomic DNA analysis, followed by hybridization with the probes in the region (shown in black on the map). The sites cleaved by a series of rare site enzymes are shown and the grouping of sites at a point indicates the presence of "HTF" islands, in all likelihood corresponding to genes.

technology and have produced good quality libraries, in some cases already extensively used. Nowadays YACs play a decisive role in the construction of the physical map for the human genome as well as for the mouse [6, 16], *Drosophila melanogaster* and even *Arabidopsis thaliana*, the generally adopted model for plant genome studies.

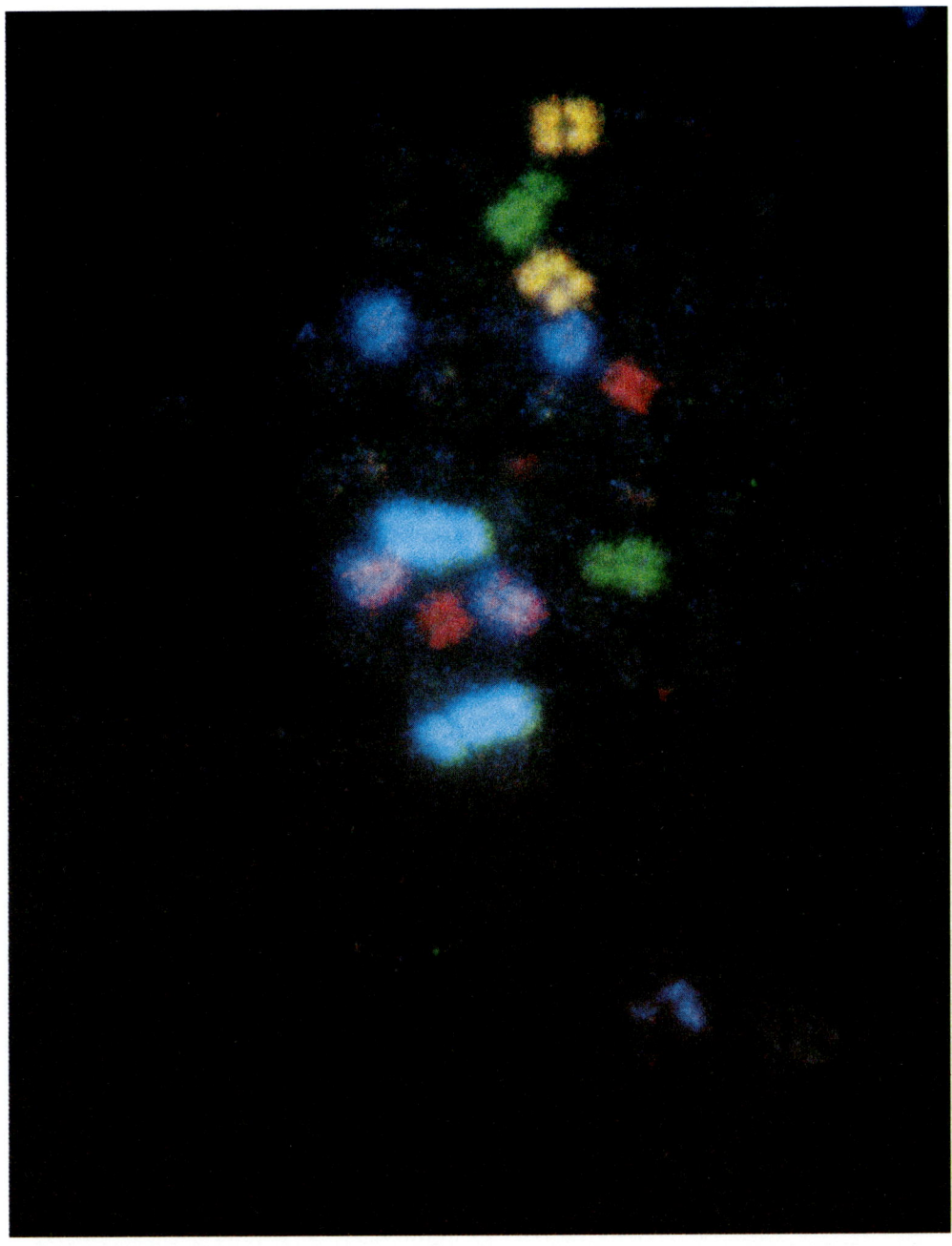

Colour plate A Chromosome painting.   This beautiful picture, provided by JH Dauwerse and GJB Van Ommen (Leyden, Netherlands), shows a human metaphase in which the various chromosomes have been "painted" with mixes of chromosome-specific probes, conjugated with different chromophores.   This technique has important applications, notably for the detection of translocations in the chromosomes of patients.

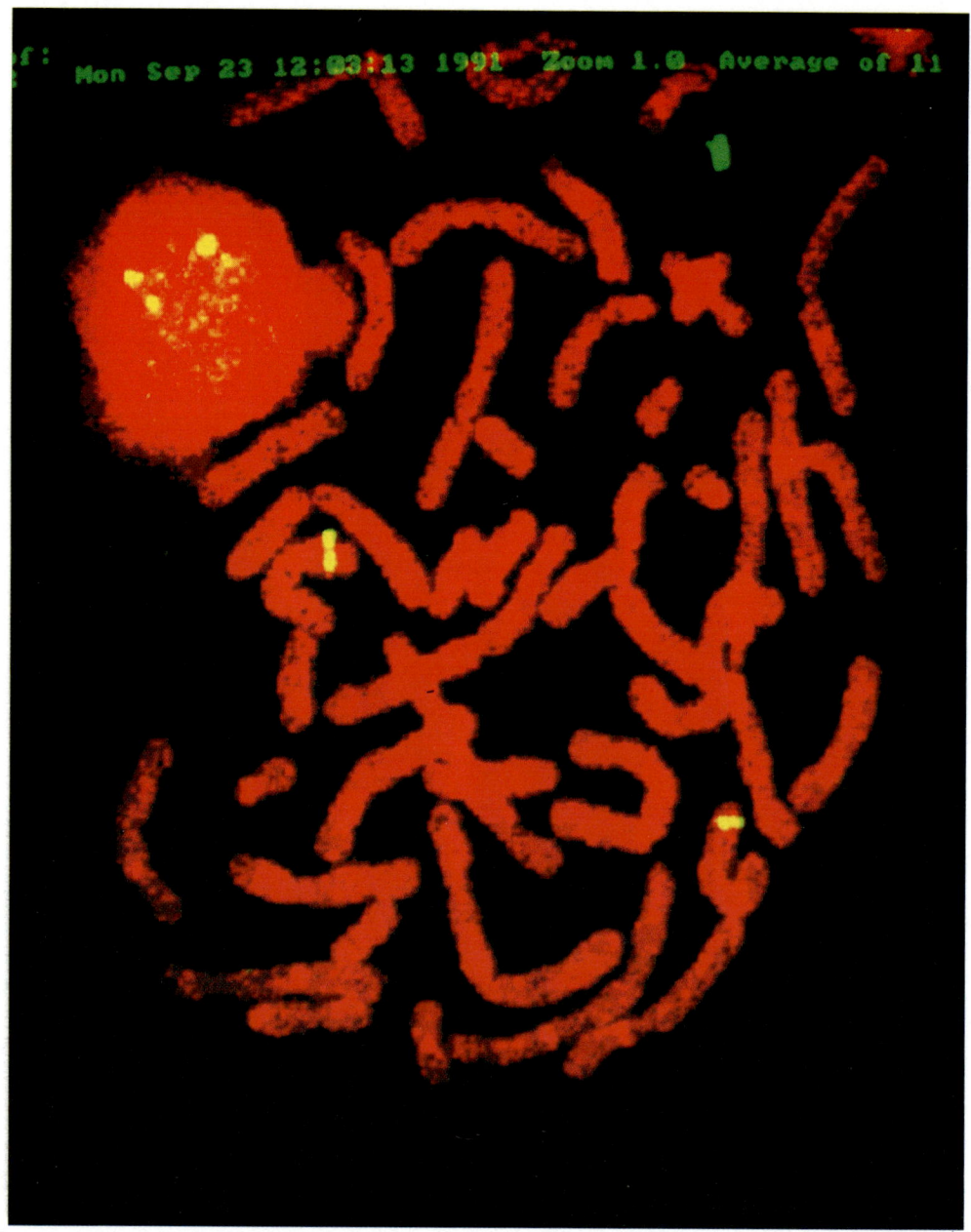

Colour plate B Localization of a gene by *in situ* **hybridization.** The probe corresponding to the gene is labelled with a fluorescent reagent and hybridizes specifically to each of the chromatids of each of the two copies of chromosome 11. (Kindly provided by Livia Selleri and Glen Evans, Salk Institute, San Diego.)

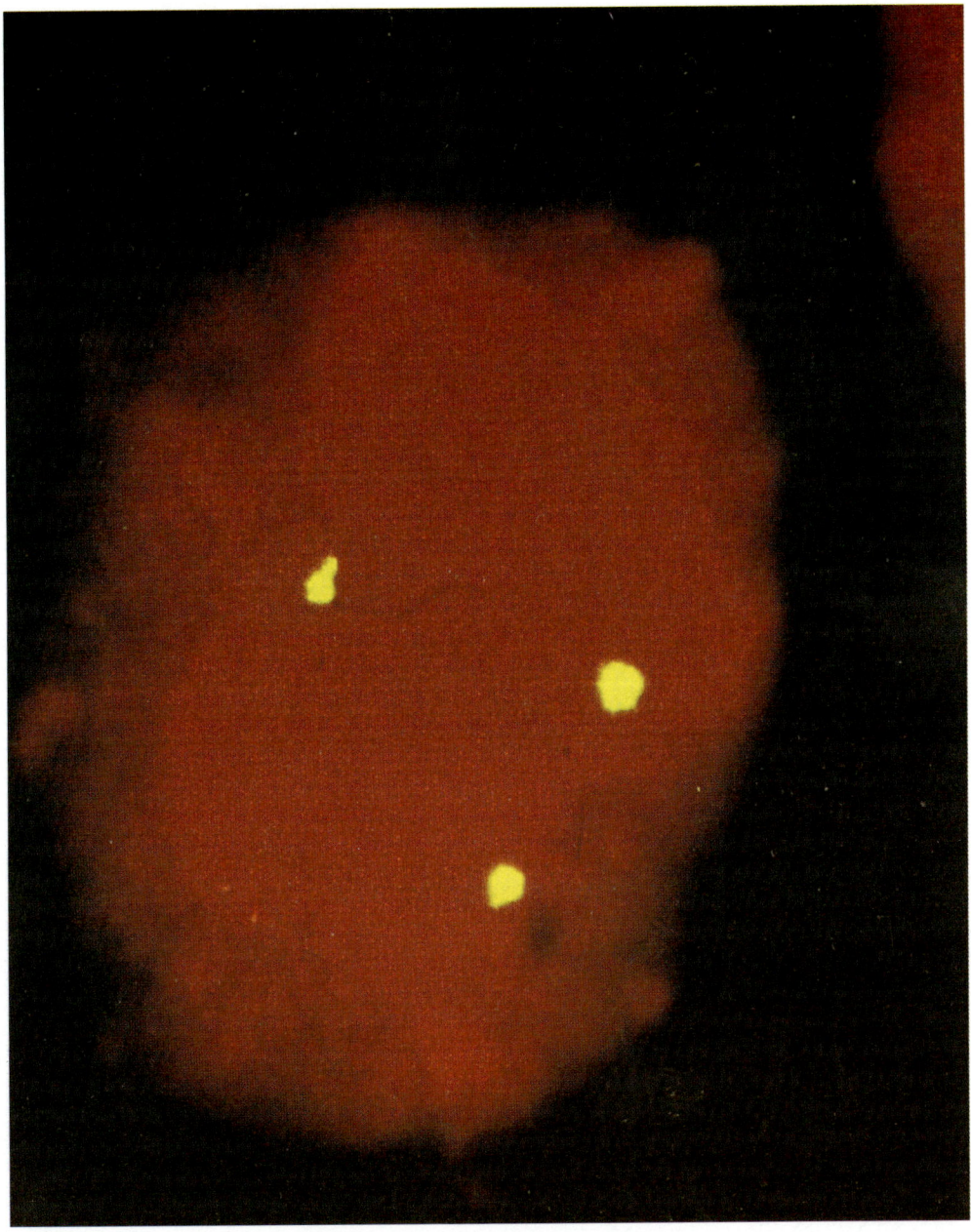

Colour plate C *In situ* **hybridization on an interphase nucleus.** As to be expected the chromosomes are not visible but the two homologous loci can be clearly distinguished. The figure shows, in fact, three rather than two bright spots because the probe contains a chromosome 21-specific sequence, and the sample originates from a Down syndrome patient. The diagnosis of trisomy is indeed one of the major clinical applications of the method. When multiple-colour probes are used, this method can also order several loci with a resolution of about a hundred kilobases.

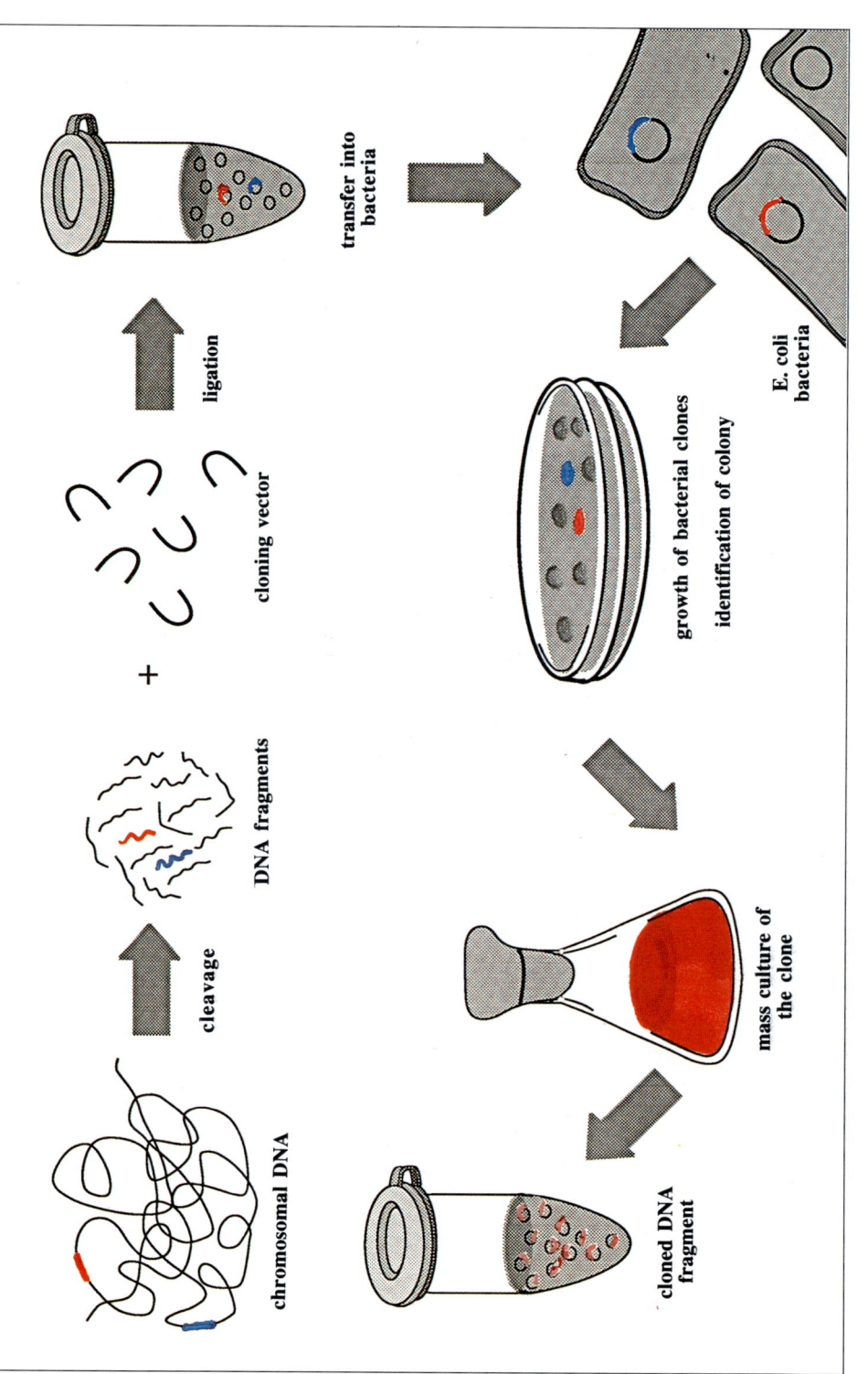

Colour plate D Cloning. This figure illustrates the steps involved in cloning a DNA segment (highlighted as a small red area in the chromosomal DNA, upper left). The DNA is cleaved to produce a mixture of fragments, each of which is ligated with a "vector" (often circular). The next steps are: introduction in bacteria; individualizing of bacterial clones, each coming from the successive divisions of a unique cell and hence containing a unique human DNA segment; identification of the colony containing the segment to be isolated (the trickiest part of cloning); and finally mass culture of the bacteria to obtain the corresponding human segment in milligram (i.e. "large") amounts.

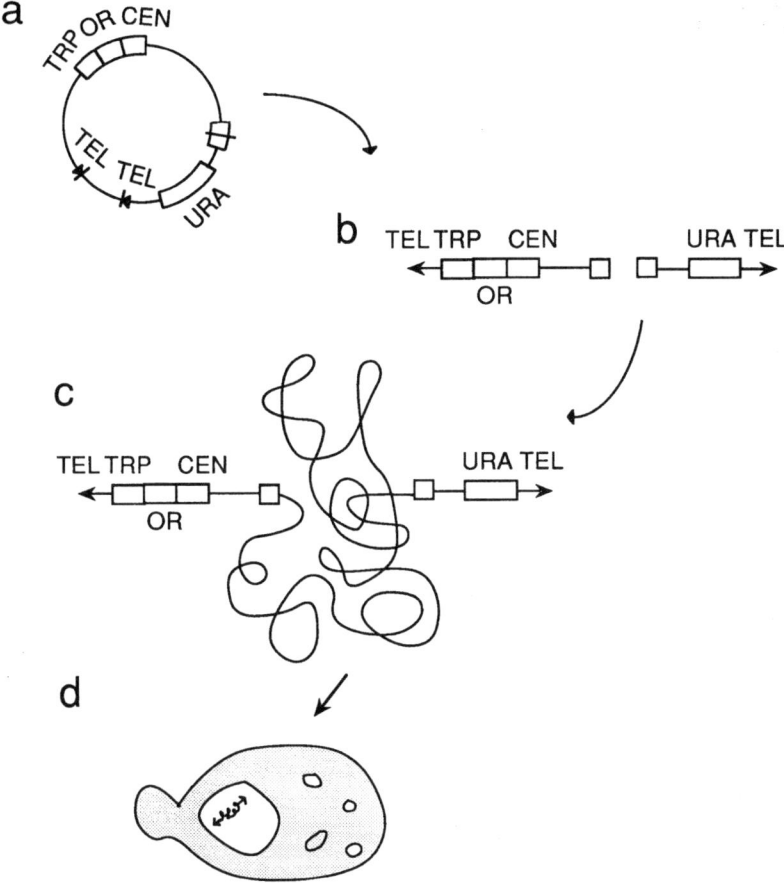

Figure 4-4 Scheme for cloning large DNA fragments as yeast artificial chromosomes (YAC). The cloning vector in circular configuration is first produced in bacteria and then cleaved to provide the two "arms" containing the telomeres (TEL), a replication origin (OR), a centromere (CEN) and the genes (URA, TRP) used for selecting the transformed yeast clones. When these arms are ligated to a very large human DNA fragment (100-1,000 kilobases) extracted from cells grown in culture, the fragment is transformed into a yeast artificial chromosome which, after introduction in yeast, is capable of replicating (thanks to the OR sequence), of stabilizing its ends (thanks to the TEL sequences) and of sharing its copies among the daughter cells during division (thanks to the CEN sequence).

A dearth of libraries

YAC technology cannot yet be considered as fully "portable" insofar as the construction of a good library is still a time-consuming and tricky task (Figure 4-5). It is not feasible, as with phages and cosmids, to produce customized libraries in just a few days, so teams wishing to use YAC clones generally apply to the laboratories

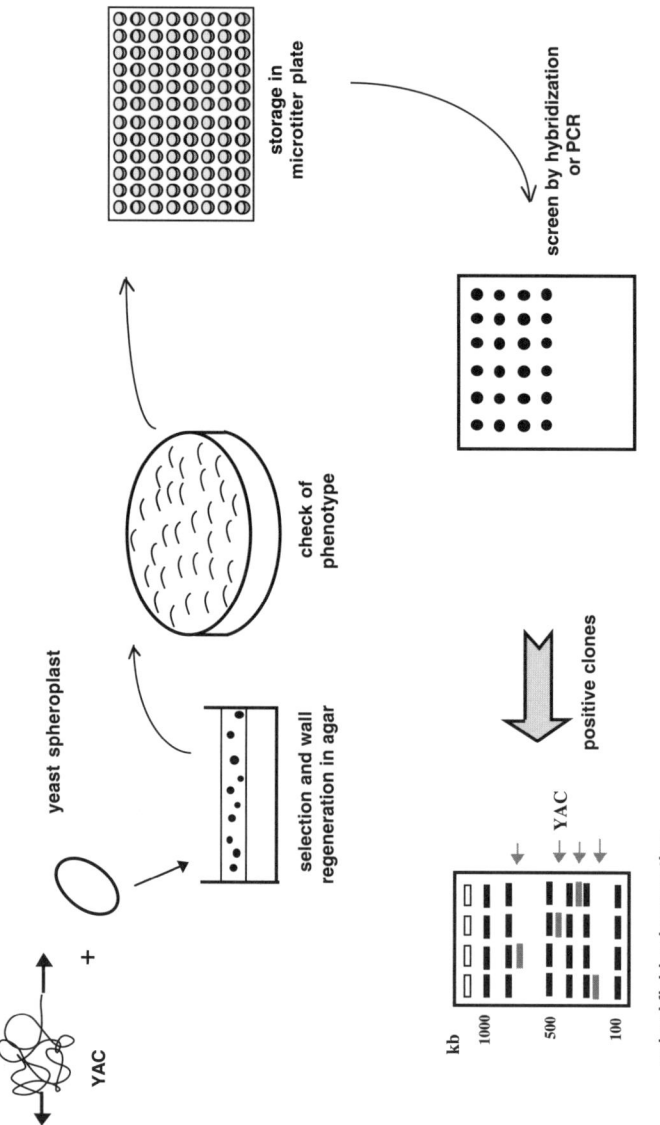

Figure 4-5 Actual steps in construction of a YAC library. First of all the large genomic DNA segments are prepared and the vector's "arms" ligated (upper left). These entities are then transferred into yeast spheroplasts (cells freed of their walls which would prevent entry of the DNA, and consequently extremely fragile). The spheroplasts are then plated in agar to regenerate their walls in the presence of a selective agent for a character conferred to the yeast cell by the YAC (auxotrophy for tryptophane). The transformation is inefficient and only a few hundred clones are transformed per microgram of DNA, compared to the usual 10^5-10^6 obtained with phages or cosmids. The clones are then manually picked onto Petri dishes with other selective agents added for checking their phenotype, and then stored in 96-well microplates. A human genome library contains roughly 50,000 clones, i.e. 500 plates. One way of using the library is to spot the clones on filters and then hybridize (lower right). The positive clones are then studied using pulse-field electrophoresis (lower left).

who have made this investment. Access is organized in various ways, usually with the help of funding from the genome programme. The most used and well known YAC collection is the Saint-Louis library set up by the groups of Maynard Olson and David Schlessinger. In fact, Saint-Louis is the home of two libraries: the first is a general one [3] containing around 60,000 clones with an average size of 300 kilobases, i.e. a six-fold coverage factor which, in theory, should guarantee that a clone will be found for almost every probe used. This general library is a first generation model with relatively small inserts and a rather large proportion of chimeric clones that contain two DNA segments from different chromosome regions. The laboratory has also set up a second library specific for the Xq24-Xq28 region [17], constructed from a human-hamster hybrid cell and containing about 800 clones with an average 200 kilobase size, and hence three times this region, estimated at about 50 megabases (50,000 kilobases). Both were initially screened by hybridization but now PCR [9] is exclusively utilized. Several hundred probes have been used for each library and the YAC clones subsequently identified have been provided to the laboratories that supplied the probes. Twelve or so copies of the Saint-Louis general library have been made and deposited in several genome centres in the USA as well as in the Resource Centre of British Genome Programme.

In Europe several teams have invested considerable time and effort in obtaining YAC collections – perhaps because access to the Saint-Louis library was more problematical than in the USA – and have obtained good results. Improvements such as DNA preparation using pulsed field gel electrophoresis (instead of sucrose gradients) and close study of the parameters of yeast transformation, have led to the creation of libraries containing inserts whose average size is about 400 kilobases (Rakesh Anand – Imperial Chemical Industries, Great Britain [2]), 500 kilobases (Hans Albertsen, Denis Le Paslier of CEPH in Paris [1]), and even 600-700 kilobases (Tony Monaco – ICRF, London [16]). CEPH has made its library widely available, and has gone on to "MegaYAC" libraries in which the average insert size is reported to exceed one megabase – a particularly efficient genome mapping tool, as demonstrated by the recent complete physical map of chromosome 21 reported by Daniel Cohen's group [7]. Such large clones do however display a number of artefacts that can make their use difficult [3].

Analysis remains tricky

Studying YAC clones is not without problems for researchers. It is clearly a major advantage to have available a very large cloned DNA segment containing a complete large gene or even the whole genome region under study. In reality, however, the detailed investigation of these clones is difficult because the very size of the cloned segments necessitates pulsed fields for analysis, and the low amount of DNA available is often a limiting factor. Whereas a segment cloned in a phage or cosmid can rather easily be produced in milligram quantities, the purification of a 400 kilobases YAC away from the 15,000 kilobases of yeast DNA present in the same cell is only possible by preparative pulsed field electrophoresis. Thus the time and effort required to prepare a few micrograms of YAC DNA are quite significant, and analysis methods have been developed to circumvent this step whenever possible. The clone can be mapped without separating it from yeast DNA through the use of end-specific probes. *In situ* non-radioactive hybridization for localizing the YAC under these conditions is also feasible although purified (or Alu-amplified) YAC DNA often gives better results. Ploys based on PCR and human-specific repeat DNA sequences (the Alu elements, usually troublesome but here put to sound use [18]) were developed so that internal probes or end probes could be obtained quickly from a YAC without purifying it away from yeast DNA (Figure 4-6).

The quicksands of chimeric clones

One as yet unsolved problem is the high percentage of "chimeric" clones found in these libraries. Chimeric clones contain two (or more) human DNA segments coming from different genomic regions and they interfere especially with physical mapping based on the construction of "contigs". The percentage of chimeric clones in the Saint-Louis library is currently estimated at 50% by its authors [11] – a sizeable proportion ! The corresponding figures for the other libraries are 20-30%, apart perhaps for the ICI library [2] with a reported 5%. Low figures for recently established libraries must be taken with a grain of salt: evaluating the percentage of these chimeric clones is not an easy task and characterizing a YAC library is both time-consuming and labour-intensive so that new libraries tend to be graced with a low figure – which however increases as time goes by and more clones are studied. Undoubtedly the best method for identifying the chimeric clones is *in situ* hybridization, with the complete YAC or preferably with its two end probes. Signals on two chromosomes or two distinct regions will indicate that the clone is, in fact, an assembly of two genomic segments.

The opinion of specialists is still divided as to the origin of these artefacts. Are they the result of coligations that took place when the library was constructed, or the consequence of subsequent recombination events ? The first theory assumes less than total efficiency both in the selection of the insert size and in the dephosphorylation treatment aimed precisely at preventing these coligations. The second conjecture is based on observation: yeast clones in YAC libraries sometimes contain two different YACs that "entered" the yeast together in the transformation phase. These two YACs can in certain circumstances recombine through pairing of repeated sequences (e.g. Alu) often found in both segments. The available data are still far too fragmentary to settle the question and, in fact, it is likely that both mechanisms play a role in the phenomenon. Regardless of the outcome, allowance must be made for chimeric inserts when using this otherwise very efficient cloning vector.

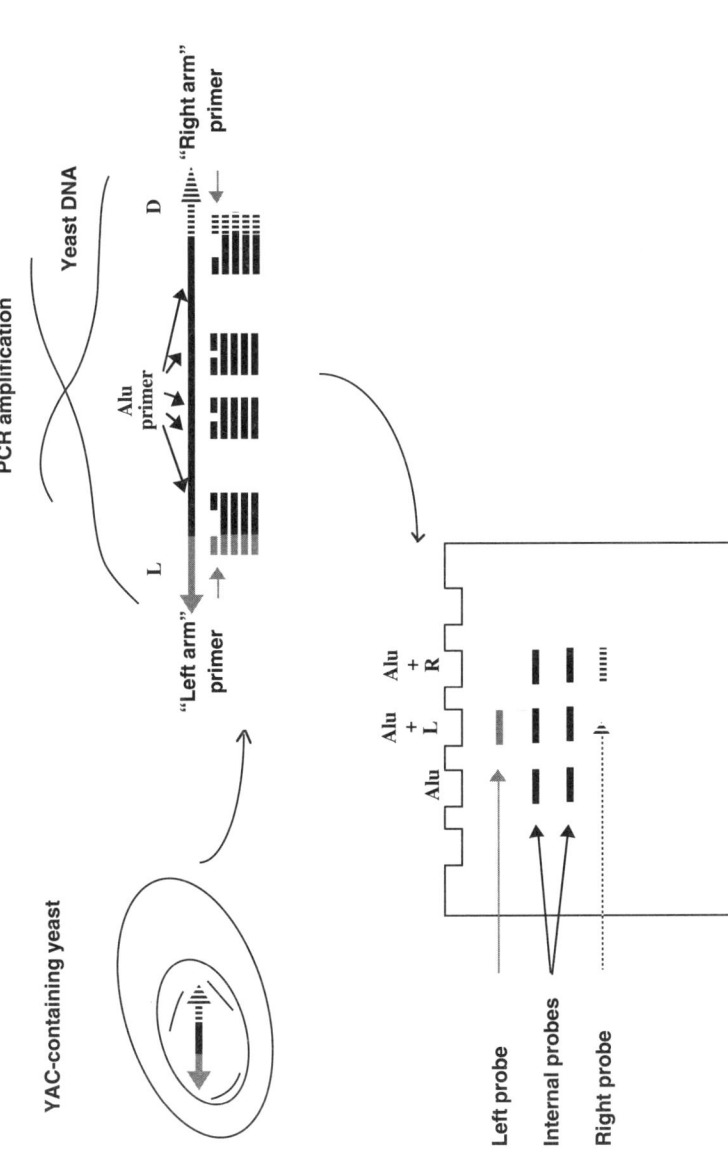

Figure 4-6 Using a YAC clone to produce internal and end probes. This process is possible without purifying the YAC clone, using PCR amplification relying on the so-called "Alu" sequences, which are species-specific and very frequent in the genome. As a general rule there are several of these sequences in every region about 100 kilobases long. PCR amplification with such primers can be performed on total DNA extracted from yeast cells containing a YAC: only those sequences belonging to a YAC and located between two Alu sequences will be amplified. The sequences corresponding to the ends of the DNA segment cloned in the YAC can also be amplified by combining Alu sequences with specific primers for each arm of the vector.

Chromosome-specific YAC libraries

A YAC library specific for a given chromosome would often have decisive advantages – it is obviously more practical to construct the physical map of a chromosome by analyzing a set of a thousand specific YACs rather than using a complete genome library with 20-30 thousand clones, the overwhelming majority of which originate from other chromosomes. Such specific collections can be created from a human/hamster or human/mouse hybrid cell containing a single human chromosome. The technique involves constructing a library from the total DNA of the hybrid and then identifying the human clones by hybridization with human-specific repeat sequences such as Alu (Figure 4-7). The Saint-Louis Xq24-Xq28 library was obtained in this way, but the method is not very satisfactory due to its inherent inefficiency – after taking considerable pains to produce the clones, more than 95% of them, those containing the hamster or mouse DNA,

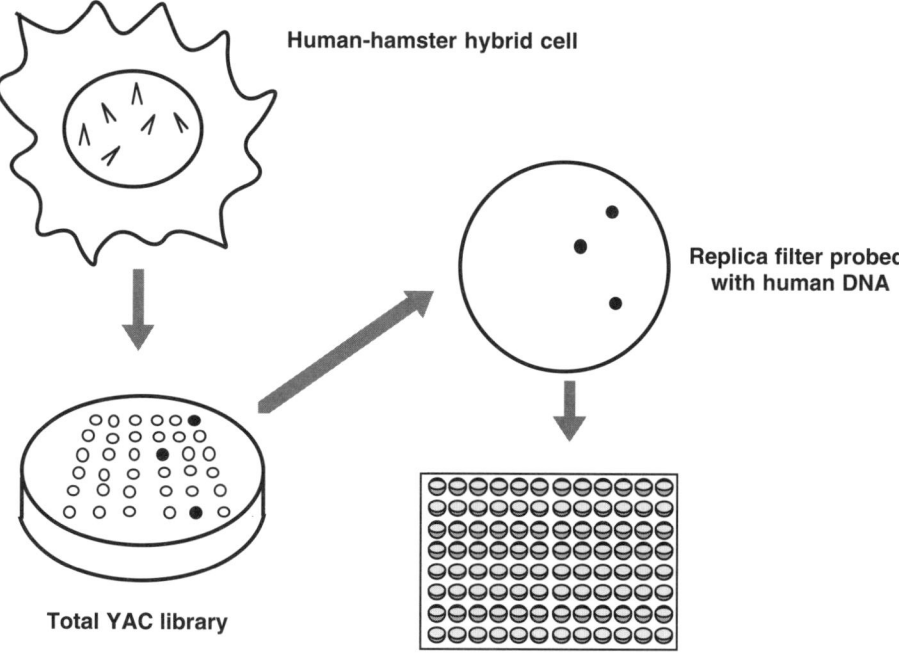

Figure 4-7 Construction of a specific YAC library for a given region of the human genome. The DNA for constructing the library comes from a human-hamster hybrid cell containing half the long arm of the human chromosome X (upper left). Total DNA from this hybrid cell is used to derive a YAC library, in which the vast majority of clones contain hamster DNA segments. Hybridization with human-specific repeat sequences identifies the clones containing human DNA inserts (i.e. 2-3% in this case), which are collected to form the Xq24-Xq28 library).

must be eliminated ! Attempts at constructing YACs from chromosomes separated in a fluorescence activated cell sorter have received close attention: they raise the hope of deriving a series of libraries for each human chromosome, as already done with phage and cosmid vectors. However the technical problems are formidable since the amount of DNA obtained from sorted chromosomes is minute, of the order of one microgram or less, and very careful tailoring of the low-yield transformation steps in YAC library construction is necessary – but not necessarily sufficient...

A third method – a very promising one – consists of "extracting" the YACs specific to a given chromosome out of a general library. It has already been established that a set of chromosome-specific probes can be used to specifically tag it by *in situ* hybridization: this is the so-called chromosome-painting procedure, now widely used (colour plate A). By analogy, it should be possible to identify in a whole-genome YAC library those clones which pertain to a particular chromosome. This approach has been used recently with success in Hans Lehrach's group in London: a complex set of probes derived from a chromosome 21-specific cosmid library has been used to select YACs from a general library, and approximately half of these turn out on further analysis to belong to the correct chromosome [20]. However, an even more efficient procedure has been published by Daniel Cohen's group [8]: in this case Alu amplification products from a chromosome 21-only hybrid cell line are used for the selection, and the resulting specific YAC library is more than 90% pure, a remarkable result. These very promising results indicate that the time may be near when chromosome-specific YAC libraries will exist – and hopefully will be made generally available, since they could tremendously speed up a number of research projects.

Reconstructing large loci

Large as they are, YACs are still not always big enough – a case in point being the giant dystrophin gene which spans three million nucleotides, while most of the YACs isolated in this region are no larger than 300 kilobases. Yet for many reasons, running from regulation studies to, possibly, gene therapy, it is often desirable to have the entire gene available in cloned form. YACs prove useful in this context because of their ability to assemble by homologous recombination in yeast. When two clones, containing a very similar or identical sequence, are inserted in the same yeast cell, they tend to recombine at this sequence. In yeast this recombination is in the vast majority of cases homologous, in contrast to mammalian cells where such an occurrence is unusual. This "building block" capability (aided and abetted by certain tricks of the trade) can thus be applied to assemble large loci, and research groups were quick to put this technique to use. The first locus reconstructed by this method was that of the cystic fibrosis gene [10]. Some researchers are even more ambitious and envisage the assembly of Mammalian Artificial Chromosomes (MAC) that could be transferred into a mammalian cell or organism and be stably propagated as an additional chromosome.

YACs in functional studies

One of the most exciting developments concerning YACs is the possibility of transfer by fusion of yeast protoplasts with mammalian cells [12]. The YAC often integrates in one or other of the cell's chromosomes without rearrangement, except at the ends, and the expression of the gene it carries can be demonstrated. This opens the way to studies on large genes transferred with their entire regulating environment. YACs can also be transferred into mice, either through pronuclear injection into fertilized oocytes [21], or by protoplast fusion with embryonic stem cells followed by reconstitution of chimaeric embryos [13]. These much-awaited results, first published in early 1993, open a number of very exciting possibilities.

All this makes YACs the preferred tool for any large-scale study on the human genome, or for that matter on other genomes: mouse [16], *Drosophila melanogaster*, *Caenorhabditis elegans* and *Arabidopsis thaliana* YAC libraries have already been set up and used extensively.

YACs are not alone

However, despite their very valuable qualities, YACs are not the perfect solution because the construction of these libraries is time-consuming and full of pitfalls, and the preparation of a reasonable amount of insert DNA is very labour-intensive – hence the tricks referred to above. There is still a need for handier, more "portable" methods that would enable a reasonably competent laboratory to prepare, with modest effort, its own library based on the somatic hybrid or the cell line most suited to its research project. Though the proven and widely used cosmid technology appears to fulfill this need, cosmid vectors have two drawbacks. The first is that the 30 to 40 kilobase size of the insert is too small: coverage of the X chromosome (150 megabases) requires over three thousand cosmids end to end. The second drawback is the instability of the DNA segments carried by these vectors. Even in the most suitable bacterial hosts, as free as possible of any recombination and repair mechanisms, a proportion of DNA sequences is found to be unstable in cosmids. These factors are a major obstacle for all mapping strategies, notably for the construction of "contigs". It is therefore not surprising that many teams are working on the development of new vectors [15].

A myriad of endeavours

The common goal of these efforts is to develop cloning systems capable of propagating large foreign DNA segments in *Escherichia coli* – preferably as a single copy to avoid recombinations but with a potential for rapid amplification when the DNA is to be prepared. Starting from a bacteriophage or a bacterial episome with a sufficiently large genome, attempts are being made to derive a practical cloning process. Historically one of the first such systems was a vector derived from bacteriophage P1 by Nat Sternberg [23]. This vector, which can in theory clone 100 kilobase DNA segments, received the backing of Du Pont de Nemours which started in 1990 to market the packaging extracts required for the construction of these libraries. However numerous laboratories attempting to use the P1 vector ran into major difficulties. A new version of the vector has been developed, and whole-genome P1 libraries have now been produced [19].

Additional candidate vectors include bacteriophage T4 which normally propagates 170 kilobases of DNA and should be capable of cloning 150 kilobase segments. An *in vitro* packaging system was defined with an apparently adequate efficiency of 10,000 clones per microgram of DNA and cloning of 95 kilobase segments was demonstrated. Nonetheless, much remains to be done before this process can be employed to construct libraries. Another candidate vector, dubbed "stealth", contains sequences from the P22 *Salmonella* phage. This system could propagate 250 kilobase segments in the amplifiable circular DNA state. It seems well designed by specialists with a sound knowledge of efficiency and of stability problems but is still in its infancy. Yet another candidate with perhaps more promise is the Bacterial Artificial Chromosome (BAC) system that employs the F factor as vector. This factor is an episome, normally present at one copy per cell in certain strains of *Escherichia coli* and that can be as large as 1,400 kilobases in circular configuration. The BAC system has already been well developed and, despite the relatively small size of the clones (i.e. up to 100 kilobases) produced in the initial tests, this approach appears to have a very bright future.

Other research groups are working on the cloning of human DNA in mammalian cells using the Epstein-Barr virus as the cloning vector, the final objective of the study being the expression more than the cloning itself. By using human cells already containing a defective virus, the amount of viral sequence necessary for propagation can be limited to 20 or so kilobases and this vector made to propagate150-180 kilobases of DNA. Sufficient quantities can then be prepared by inducing a lytic cycle during which numerous viral particles containing the cloned DNA segment will be generated. This process is attractive, all the more so because it is very suitable for studying the expression of genes carried on the cloned segment and because gene identification could be achieved by complementation of a function, as in yeast. Yet another prospect is the cloning of 1-10 megabase segments as minichromosomes (double minute) in mouse cells. The principle of the method is to introduce beforehand in the region to be cloned a neomycin-resistance gene linked with a dihydrofolate reductase (DHFR) gene. After irradiation, fusion with the mouse cell and selection with neomycin, amplification of the inserted segment can be triggered by cell culture in the presence of methotrexate – the methotrexate/DHFR system is the very first one for which gene amplification was actually demonstrated. Logically the next step is to separate these double minutes, possibly using pulse-field gel electrophoresis.

It would be clearly foolhardy at this time to predict which of these methods will prove appropriate and have general applicability. What can be said is that the basic concepts have been validated in most cases – but there is still a long way to go before they emerge as competitive cloning methods. It should, however, be clear from this overview that molecular biology technologies are still evolving rapidly and that a major breakthrough (as, in its time, the development of the YAC system) can still occur.

REFERENCES

1. Albertsen HM, Abderrahim H, Cann HM, Dausset J, Le Paslier D, Cohen D: Construction and characterization of a yeast artificial chromosome library containing seven haploid Human Genome equivalents. *Proc Natl Acad Sci USA* 1990 **87:** 4256-4260

2. Anand R, Riley JH, Smith JC, Markham AF: A 3.5 genome equivalent multi access YAC library: construction, characterisation, screening and storage. *Nucleic Acids Res* 1990 **18:** 1951-1956

3. Anderson C: Genome shortcut leads to problems. *Science* 1993 **259:** 1684-1687

4. Brownstein BH, Silverman GA, Little RD, Burke DT, Korsmeyer SJ, Schlessinger D, Olson MV: Isolation of single-copy human genes from a library of yeast artificial chromosome clones. *Science* 1989 **244:** 1348-1351

5. Burke DT, Carle GF, Olson MV: Cloning of large segments of exogenous DNA into yeast by means of artificial chromosome vectors. *Science* 1987 **236:** 806-812

6. Chartier FL, Keer JT, Sutcliffe MJ, Henriques DA, Mileham P et al: Construction of a mouse yeast artificial chromosome library in a recombination-deficient strain of yeast. *Nature Genetics* 1992 **1:** 132-136

7. Chumakov I, Rigault P, Guillou S, Ougen P, Billault A et al: A continuum of overlapping clones spanning the entire human chromosome 21q. *Nature* 1992 **359:** 380-387

8. Chumakov IM, Le Gall I, Billault A, Ougen P, Soularue P et al: Isolation of chromosome 21-specific yeast artificial chromosomes from a total human genome library. *Nature Genetics* 1992 **1:** 222-225

9. Green ED, Olson M: Systematic screening of yeast artificial chromosome libraries by use of the polymerase chain reaction. *Proc Natl Acad Sci USA* 1990 **87:** 1213-1217

10. Green ED, Olson M: Chromosomal region of the cystic fibrosis gene in yeast artificial chromosomes: a model for human genome mapping. *Science* 1990 **250:** 94-98

11. Green ED, Riethman HC, Dutchik JE, Olson MV: Detection and characterization of chimeric yeast artificial chromosome clones. *Genomics* 1991 **11:** 658-669

12. Huxley C, Gnirke A: Transfer of yeast artificial chromosomes from yeast to mammalian cells. *Bioessays* 1991 **13:** 545-550
13. Jacobovits AJ, Moore AL, Green LL, Vergara GJ, Maynard-Currie CE, Austin HA, Klapholz S: Germ-line transmission and expression of a human-derived yeast artificial chromosome. *Nature* 1993 **362:** 255-258
14. Jordan BR: YAC Power. *Bioessays* 1990 **12:** 183-187
15. Jordan BR: Des vecteurs de clonage à la pelle. *Médecine/Sciences* 1991 **7:** 503-504
16. Larin Z, Monaco AP, Lehrach H: Yeast artificial chromosome libraries containing large inserts from mouse and human DNA. *Proc Natl Acad Sci USA* 1991 **88:** 4123-4127
17. Litlle RD, Porta G, Carle GF, Schlessinger D, D'Urso M: Yeast artificial chromosomes with 200- to 800-kilobases inserts of human DNA containing HLA, V. kappa, 5S, and Xq24-Xq28 sequences. *Proc Natl Acad Sci USA* 1989 **86:** 1598-1602
18. Nelson DL, Ledbetter SA, Corbo L, Victoria MF, Ramirez-Solis R, Webster TD, Ledbetter DH, Caskey CT: Alu polymerase chain reaction: a method for rapid isolation of human-specific sequences from complex DNA sources. *Proc Natl Acad Sci USA* 1989 **86:** 6686-6690
19. Pierce JC, Sternberg N, Sauer B: A mouse genomic library in the bacteriophage P1 system: organization and characterization. *Mammalian Genome* 1992 **3:** 550-558
20. Ross MT, Nizetic D, Nguyen C, Knights C, Vatcheva R et al: Selection of a human chromosome 21 enriched YAC sub-library using a chromosome-specific composite probe. *Nature Genetics* 1992 **1:** 284-290
21. Schedl A, Montoliu L, Kelsey G, Schütz G: A yeast artificial chromosome covering the tyrosinase gene confers copy number-dependent expression in transgenic mice. *Nature* 1993 **362:** 258-261
22. Schwartz DC, Cantor CR: Separation of yeast chromosome-sized DNAs by pulsed field gel electrophoresis. *Cell* 1984 **37:** 67-75
23. Sternberg N: Bacteriophage P1 cloning system for the isolation, amplification and recovery of DNA fragments as large as 100 kilobase pairs. *Proc Natl Acad Sci USA* 1990 **87:** 103-107

5
Strategies for an integrated map

Physical maps come in several varieties

So far we have not discussed the different implementations of a physical map. It may be based mainly on analysis of genomic DNA (e.g. in pulse-field gels), and will then indicate the size of the region, the positions of a number of landmarks such as probes and restriction sites. However, such a map does not provide quick access to cloned DNA for every point of the interval. On the other hand, if the map has been constructed by "covering" the region with cloned DNA segments, it will provide not only such topographical data but also the underlying pieces of cloned DNA, each of which can be prepared in large quantity for further study (Figure 5-1). This will open up the possibility of studying any specific area in detail, mapping it very precisely, searching for genes and even sequencing all or part of it. A bona fide and fully operational physical map thus consists of a line-up of contiguous, overlapping clones, a "contig", extending across the entire region under study, as was achieved in 1987 for the whole genome of the *Escherichia coli* bacterium (Figure 2-3, p. 10), which spans a "mere" 5,000 kilobases [12]. Various laboratories are currently attempting to construct such maps for human chromosomes, with recent success for chromosome 21 [6, 7] and the Y chromosome [8].

Such mapping exercises start with hundreds or thousands of clones (cosmids or, preferably, YACs), which correspond to a given region of a chromosome or even to a complete chromosome, and are analyzed and compared with each other to determine which of them overlap. Clones can be "assembled" if overlaps have been identified; the catchword used by researchers for such a group of contiguous clones is a "contig". Finding these contigs determines – by definition – the physical map of the area they cover. In addition, the region

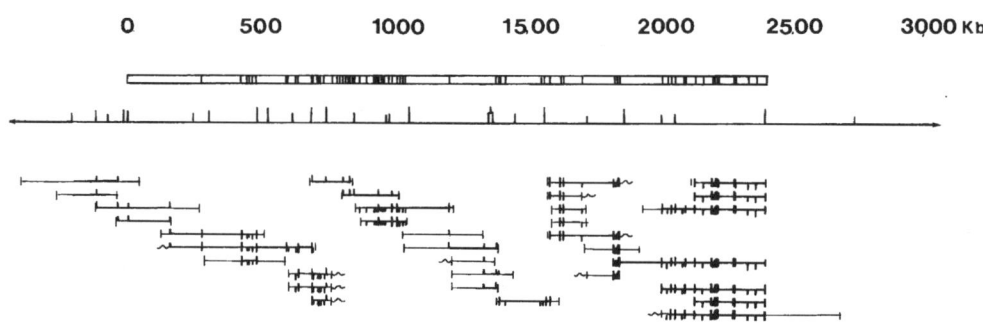

Figure 5-1 The different types of physical maps. This figure contrasts – in a somewhat extreme and schematic fashion – a physical map constructed using pulsed-field analysis of genomic DNA (with several probes in the region to detect the respective fragments), with a map based on cloned DNA analysis and assembly of the clones into contigs. In the top half, a 1988 restriction map of the dystrophin gene region shows the sites cleaved by the SfiI enzyme and the size of the cleaved fragments. The probes used to construct the map are symbolized by the black squares across the top. They were the only cloned DNA segments existing in the region and though the analysis reveals the area's anatomy, it does not provide direct access to the other points of the map. In the lower half, a map of the same region published recently by a British group led by David Bentley. From top to bottom, the lower diagram shows : at the top, the dystrophin gene and its various exons (in black); in the middle, a map with a number of precisely positioned probes; and at the bottom, the YACs used to derive the map. The vertical bars on the YACs indicate the location of the probes (in fact STS) used to order them. The wavy lines at the ends of some YACs symbolize a DNA segment coming from another region (chimeric clones). This map is much more accurate and reliable than the first; in addition the YAC clones make any point accessible for further study, in as much detail as required, even down to the sequence.

then becomes directly accessible as it is broken down into small ordered segments, each of which is an existing clone, tucked away in a given well of a numbered 96-well plate in a given deep-freezer. If genetic analysis indicates that the unknown gene implicated in a disease is located between probes A and B and if a contig of clones extending from A to B has been set up, then one of these clones necessarily contains the gene. It will then be possible to identify it by more detailed study: a catalogue of the genes contained in these segments can be established, and each of them compared in patients and controls. A gene that is systematically deleted or mutated in affected persons is very likely to be the gene involved in the disease.

From region... to chromosome

Physical mapping may focus on a limited area, one or several megabases separating two genetic markers that span the locus of the disease being studied: in this case, it is a targeted strategy whose aim is to identify and isolate this gene as quickly as possible, according to the now classical approach of reverse genetics (or "positional cloning"). This method has been applied to isolate the dystrophin gene involved in Duchenne muscular dystrophy [5], the cystic fibrosis gene [18], and many others. The gene causing Huntington's chorea, whose location was approximately determined in 1983 [9], was finally identified in 1993 [21] after ten years of strenuous efforts.

On the other hand, laboratories participating in genome programme study the whole genome, or one chromosome, or at least one chromosome arm, with the aim of constructing its complete physical map and assembling collections of clones spanning the entire region, that will be extremely valuable for any subsequent study of a gene or a disease localized in any particular region. There are very significant advantages inherent in doing things this way. If a laboratory focuses on the exhaustive study of one chromosome, it can make use of existing specific libraries as the starting point. All the resulting data and all the clones analyzed will then be part and parcel of the overall project. Furthermore, from one end of the chromosome to the other, the work will be of uniform quality and the efforts will be better coordinated than attempts to patch together results coming from small groups working separately with different technologies and little overall cohesion. This facet of genome programmes is in fact well suited to a "chromosome-by-chromosome" approach, as evidenced by the results of the last two or three years that clearly underscore the inherent advantages of such a systematic approach. More daring whole-genome mapping approaches [2] may, however, complement or supersede these efforts.

Chromosome 19 at Lawrence Livermore

A few figures will give a feeling for what this work involves. Our reference will be one of the DOE laboratories in California – the Lawrence Livermore Genome Center directed by Anthony Carrano (Figure 5-2), whose aim is to construct the physical map of chromosome 19, which spans approximately 70 megabases. The method consists of analyzing and assembling cosmids from a specific chromosome 19 library. The cosmids are cleaved with enzymes and the ends tagged by a fluorescent compound: each of them yields a few dozen fragments that define its signature. The fragment sets are analyzed semi-automatically and then compared pairwise to detect any similarities indicating possible overlap [22, 23]. In three years, over 8,000 clones have been studied and the resulting contigs, whose average size is around 100-150 kilobases, cover more than 60% of the chromosome, although there are still many "gaps" between them. However, a YAC library is being used to help define the junctions, and *in situ* hybridization confirms whether the assemblies are valid and locates them along the chromosome. These activities require intensive use of data processing, relatively sophisticated equipment and a staff of about 40 persons. The culmination of the physical map, i.e. a virtually

Figure 5-2 Tony Carrano, Director of the Lawrence Livermore Genome Center (Department of Energy) in California. (Photograph taken at Cold Spring Harbor.)

complete series of cosmids and YAC contigs aligned along the chromosome, is now in sight. Though labour-intensive, resource-intensive and systematic, this programme is now on the verge of reaching its objective, despite the scepticism reigning at its inception. Without any doubt YACs were discovered at just the right time to compensate for the limitations of cosmids!

In fact, whole-chromosome, or even whole-genome, mapping strategies directly based on YACs can be devised, and their implementation has shown them to be very effective. Cases in point are the chromosome 7 and X mapping efforts in Saint-Louis, the YAC fingerprinting procedure set up at Généthon in France [3, 7] and the Y chromosome mapping effort in Cambridge (USA) [8]. The cosmid contig approaches used at Livermore (and, in a somewhat different manner, at Los Alamos [19]) suffer from the fact that they were devised in the late eighties – at the time when YACs were not yet established as a possible alternative.

Have sequence tagged sites won the day ?

The concept

Physical mapping projects are subject to a multitude of occasionally contradictory constraints such as speed, reliability, suitability of the data and objects produced for use by external laboratories, and affordable cost. Thus widely different strategies and organizational methods are implemented, yet all these approaches have to be reconciled and unified at some point. The concept of Sequence Tagged Sites (STS), published by Olson et al [15] in late 1989, is an attempt to systematize landmarking of the human genome. The STS system has gained general acceptance throughout the USA to the point that some take it for the gospel truth – while others, especially the "reference library" champions in Great Britain (see below), believe that this is not by any means the most efficient approach.

Let us first describe the concept. It involves determining a few hundred bases of DNA sequence at a large number of points in the genome. For each of these locations a specific couple of oligonucleotides is then defined. PCR amplification on a sample of total human DNA, with these two oligonucleotides as primers, produces a unique segment of DNA, usually several hundred bases long. Together all the STS form a chain of landmarks for the physical map (and possibly for the genetic map). The main originality of this concept is that the STS is specified by the sequence of the two primers that make its production possible. Thus it is defined once and for all, regardless of whether, for instance, the region under study is cloned in a phage, cosmid or YAC. Moreover, the STS will remain valid should the corresponding area of the genome be cloned sometime in the future in a new, as yet unknown vector. The STS is by definition fully portable since it can be obtained once the sequences of its two oligonucleotides are known: one or two hours on a PCR machine and a small quantity of human genomic DNA is all that is required to produce a probe for use in any desired study. One of the major advantages of STS is that the time-consuming collection of probes from the many researchers responsible for their isolation (or even from the American Type Culture Collection) is replaced by simple look-up in a data bank followed by quick (if somewhat expensive) local production of oligonucleotide pairs. A further advantage is that the definition of STS can easily integrate data coming from different laboratories, since a simple PCR reaction can determine whether a given YAC or cosmid contains a specific STS. This was the meaning of Olson et al, when they advocated STS as "a common language for physical mapping of the human genome" [15].

Where are we today ?

What is the situation three years after this seemingly infallible system came into use ? The takeover by STS has only partially succeeded; their widespread use has run into several difficulties. The first practical problem is the cost of the oligonucleotides. It is true that many laboratories are now equipped with oligonucleotide synthesizers and, for those without, a quick glance through the pages of *Science* or *Nature* emphasizes the fierce competition waged by suppliers by means of discount wars and inflated promises on the quick delivery and high quality of these reagents. Notwithstanding, it still costs one or two hundred dollars to synthesize or purchase the oligonucleotide pair needed to characterize an STS; thus the cost for a research team implementing a hundred or so STS (the minimum requirement for studying the map of a reasonably sized region) is significant. In addition, this money is largely wasted since current synthesizers produce one hundred, if not one thousand times, more product than is required for an amplification reaction. The teams that have prepared the primers for an STS set are therefore now sharing them with their collaborating groups. The result is that though probes are no longer sent by mail, they have been replaced by packages containing oligonucleotides... Moreover the portability of these reagents is not always guaranteed, as it is conditional on amplification conditions being perfectly reproducible in any laboratory. Practical experience shows that primers that "work" (i.e. significantly amplify a well defined fragment from genomic DNA) in one laboratory sometimes fail in another. Apart from quirks in buffer makeup, water quality and sources of reagents, the main reason for this variability lies with the PCR "machines". In this respect, a study [10] - harshly critical of some suppliers - found that the temperature profiles produced by some thermal cyclers bears little resemblance to the advertised curve. Thus the "reagent" formed by these two sequences of 20 or so nucleotides must often undergo development before it can be used, thereby making its implementation less appealing.

The second difficulty lies in the principle: while STS are theoretically excellent for physical mapping (as the title of the paper [15] by Olson et al indicates), they are not *a priori* designed to connect this approach with genetic mapping. Yet this linkage is of paramount importance, as proven by recent events, such as the problems encountered with Huntington's chorea [17] or the progress made on the fragile X syndrome [4, 11]. Far from removing the need for genetic mapping, large scale DNA studies employing the most advanced techniques urgently need landmarks spaced every few megabases, that currently only genetic maps are capable of supplying and ordering. The grid proposed by Olson and coworkers, with one STS every 100 kilobases, would be immensely more efficient if at least some of these markers were to feature polymorphism making them also operational in this respect. It appears that this aspect had not been included in the STS proposal, at least as it was initially formulated.

STS, EST, microsatellites... Towards the perfect landmark ?

Nowadays STS are often combined with microsatellites [20], the almost perfect markers for genetic studies. Microsatellites are analyzed by PCR amplification with oligonucleotides that recognize the single copy sequences surrounding them, followed by electrophoresis on extremely high resolution acrylamide gel to detect their length variations (Figure 3-3, p. 25). The beauty of this process is that it generates a "super STS" each time. The result is well and truly an STS because production of this segment of the genome merely requires that the sequence of the oligonucleotides (and the required PCR conditions) be known; however, this STS is a "top of the line" model because it also serves as a polymorphic marker in genetic analysis. In another connection, massive (but partial) sequencing of complementary DNA clones also produces markers, already named Expressed Sequence Tags (EST) [1] by some authors. It should be stressed that, though many of these super STS have currently been defined, too few of them have been released to the scientific community – as emphasized by heads of data banks [16]. This regrettable situation results from the ongoing and sometimes bitter competition to study certain "hot" areas of the human genome. The public or private decision-makers responsible for allocating the substantial sums devoted to this type of research will have to find suitable and perhaps coercive schemes to ensure that the whole scientific community does indeed benefit from the reagents produced by means of genome funding...

Another approach: reference libraries

The construction of physical maps can be organized in more than one way. One of the most original approaches has been devised by Hans Lehrach, a scientist of Austrian origin who heads a group of 30 or so persons at the Imperial Cancer Research Fund (ICRF) laboratories in London. His approach completely reverses the STS scheme in that it is based on libraries of DNA cloned in cosmids, YACs or other vectors, kept as collections of individual clones in microtiter dishes. As shown in Figure 5-3, the method involves using a suitable robot to spot the clones of the library as a regular and reproducible array on a filter, ten or twenty thousand at a time. Batches of filters can then be distributed outside for screening by each laboratory with its pet probes: a few standard-sized filters contain an entire library. The address of the corresponding clones (i.e. number of plate, number of well) is determined by the location on the filter of the positive signals obtained, since the arrangement of the clones on the membrane is known and permanent. The London laboratory can then provide positive clones to the outside group without having to transfer the library itself, while at the same time having acquired the information about which clone hybridizes with which probe. Thus the data is quite naturally centralized and accumulated. This shared approach to using the bank can be very useful to outside laboratories. But it nevertheless guarantees data centralization and ultimately construction of maps over large distances, although this can only be achieved by performing some additional experiments, such as hybridizing a moderate number of repeat sequences on the filters. These sequences define the "signature" of each of the clones and indicate their possible overlaps.

This strategy is thus built around the libraries, their distribution to numerous collaborators and data generation through hybridization. Clearly this strategy is quite opposed to the concepts prevailing in the USA, where STS rather than clones are the reference objects and PCR amplification rather than hybridization – not considered reproducible enough – is selected for screening. Hybridization does have the overwhelming advantage of being suitable for highly parallel approaches that maximize the volume of information per experiment. Hans Lehrach has published several very persuasive papers on this subject [13]. Only time will tell that of these two strategies (there are others as well) will ultimately prove to be the most effective; the complete link-up of the Schizosaccharomyces pombe genome by the London group [14] shows the power of their method. In any event the appeal of the reference libraries in the short term is that the other laboratories can easily access the clones of interest to them, particularly the YACs. Alternate mapping strategies, such as the cosmid contig approach pursued in some US Genome Centers, produce fewer useful intermediate results for outside users before the final map is obtained.

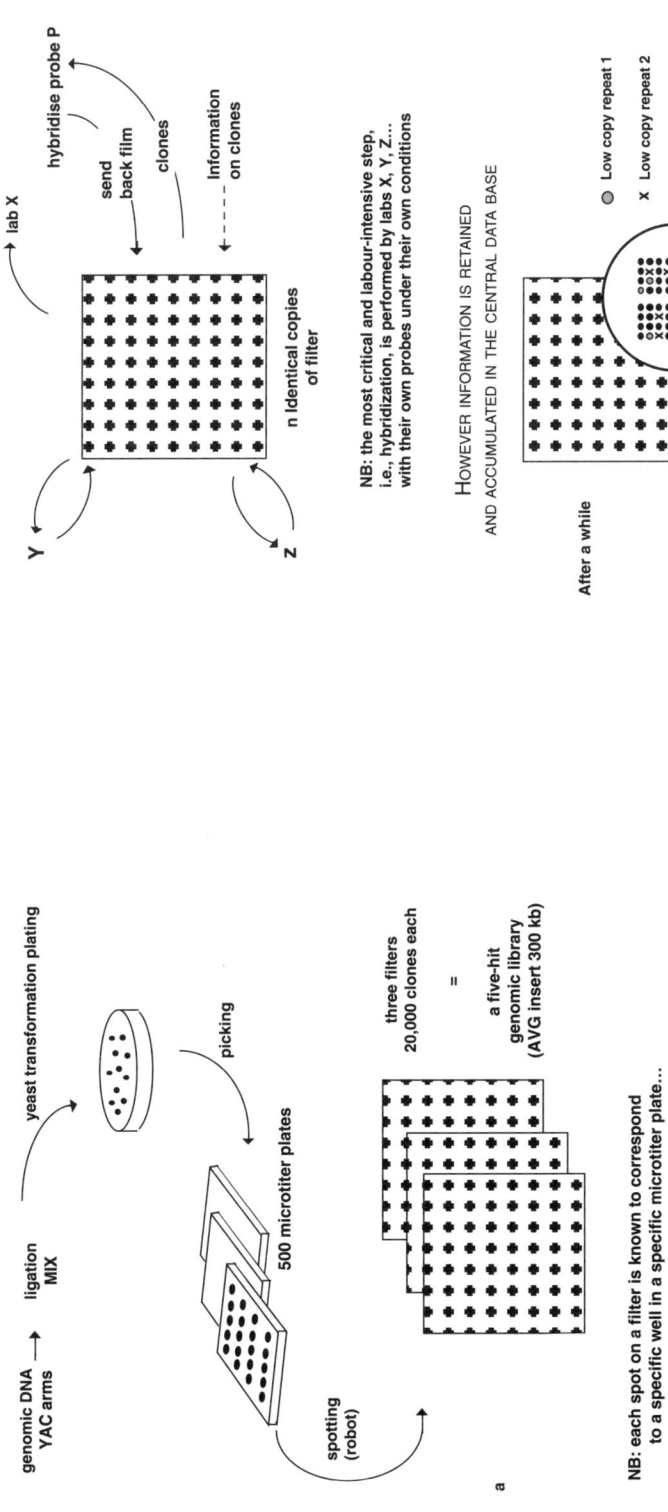

Figure 5-3 The concept of the reference library. A YAC library containing about 50,000 clones stored in 500 microtiter plates is shown on the left. Each of these clones is individually and very accurately positioned on the filters in a very fine grid (20,000 clones per filter) so that each position corresponds to a defined well in a particular microtiter plate. The procedure for utilizing the reference library is shown on the right: sets of identical filters are produced and sent to the laboratories wishing to isolate the YAC(s) corresponding to "their" probe. These laboratories then hybridize the filters and notify the central laboratory of the position of the positive clone(s), which can then be provided. The lower half of the figure shows how information on each of the library's clones progressively and naturally builds up.

Simultaneous YAC contig building over the whole genome: a surprise from Généthon

The most daring approach, however, is certainly that implemented by Daniel Cohen's group (CEPH and Généthon, France), who set out to obtain and correlate "signatures" from all clones of a "MegaYAC" library, so as to cover the whole human genome with contigs based on these clones. The method involves digesting each YAC with several restriction enzymes, running the fragments on gels and blotting them (using the automated instruments installed at Généthon), hybridizing with human repeated sequences to reveal several bands for each YAC and comparing each of these signatures to each other using a network of powerful computers. The various steps have been implemented in a very controlled way, and results obtained by summer 1992 indicated that approximately half of the genome was already covered in contigs of large size, more than two megabases [3]. In other words, this very successful implementation of a YAC fingerprinting method appeared close to providing a general framework physical map of the whole human genome, a result that was not expected before several years. This rapid advance did not go unnoticed in the USA; it has prompted a reorientation of the dominant chromosome-by-chromosome strategy, and motivated rapid distribution of the CEPH "MegaYAC" library at the end of 1992. Since then, some doubts and criticisms have been voiced, centred on the frequency of chimeric clones and the occurrence of deletions and rearrangements in some of the YACs [2]. It may be that too much was expected too soon ; some US critics may also have an axe to grind in the process... In addition, it is clear that judging the quality of a YAC library, or the accuracy of a large physical map, is even more lengthy and difficult than the construction of these entities. A reliable assessment of this achievement will be forthcoming through the work of the numerous groups that are either using or refining it. The importance of this breakthrough, however, is undeniable.

REFERENCES

1. Adams MD, Kelley JM, Gocayne JD, Dubnick M, Polymeropoulos MH et al: Complementary DNA sequencing: expressed sequence tags and Human Genome project. *Science* 1991 **252**: 1651-1656
2. Anderson C: Genome shortcut leads to problems. *Science* 1993 **259**: 1684-1687
3. Bellanne-Chantelot C, Lacroix B, Ougen P, Billault A, Beaufils S et al: Mapping the whole human genome by fingerprinting yeast artificial chromosomes. *Cell* 1992 **70**: 1059-1067
4. Brown WT: The fragile X: progress toward solving the puzzle. *Annu J Hum Genet* 1990 **47**: 175-180
5. Chelly J: La myopathie de Duchenne: du gène DMD à la dystrophine. *Médecine/Sciences* 1988 **4**: 141-150

6. Chumakov IM, Le Gall I, Billault A, Ougen P, Soularue P et al: Isolation of chromosome 21-specific yeast artificial chromosomes from a total human genome library. *Nature Genetics* 1992 **1:** 222-225
7. Chumakov I, Rigault P, Guillou S, Ougen P, Billaut A et al: A continuum of overlapping clones spanning the entire human chromosome 21q. *Nature* 1992, **359:** 380-387
8. Foote S, Vollrath D, Hilton A, Page DC: The human Y chromosome. Overlapping DNA clones spanning the euchromatic region. *Science* 1992 **258:** 50-66
9. Gusella JF, Wexler NS, Conneally PM, Naylor SL, Anderson MA et al: A polymorphic DNA marker genetically linked to Huntington's disease. *Nature* 1983 **306:** 234-238
10. Hoelzel R: The trouble with PCR machines. *Trends Genet* 1990 **6:** 237-238
11. Jordan B: Fragile X-linked mental retardation and the difficulties of reverse genetics. *Bioessays* 1991 **13:** 243-251
12. Kohara Y, Akiyama K, Isono K: The physical map of the whole E. coli chromosome. *Cell* 1987 **50:** 495-508
13. Lehrach H: Hybridization fingerprinting in genome mapping and sequencing. *Genome Analysis* **1:** 39-81; Cold Spring Harbor Laboratory Press, 1990
14. Maier E, Hoheisel JD, McCarthy L, Mott R, Grigoriev AV: Complete coverage of the schizosaccharomyces pombe genome in yeast artificial chromosomes. *Nature Genetics* 1992 **1:** 273-277
15. Olson M, Hood L, Cantor C, Botstein D: A common language for physical mapping of the Human Genome. *Science* 1989 **245:** 1434-1435
16. Pearson PL, Maidak B, Chipperfield M, Robbins R: The Human Genome initiative – Do databases reflect current progress? (Perspective) *Science* 1991 **254:** 214
17. Roberts L: Huntington's gene: so near, yet so far. *Science* 1990 **247:** 624-627
18. Rommens JM, Iannuzzi MC, Kerem B, Drumm ML, Melmer G et al: Identification of the cystic fibrosis gene: chromosome walking and jumping. *Science* 1989 **245:** 1059-1065
19. Stallings RL, Doggett NA, Callen D, Apostolou S, Chen et al: Evaluation of a cosmid contig physical map of human chromosome 16. *Genomics* 1992 **13:** 1031-1039
20. Tautz D: Hypervariability of simple sequences as a general source for polymorphic DNA markers. *Nucleic Acids Res* 1989 **17:** 6463-6471
21. The Huntington's disease collaborative research group: A novel gene containing a trinucleotide repeat that is expanded and unstable on Huntington's disease chromosomes. *Cell* 1993 **72:** 971-983
22. Trask B, Christensen M, Fertitta A, Bergmann A, Ashworth L et al: Fluorescence in situ hybridization mapping of human chromosome 19: Mapping and verification of cosmid contigs formed by random restriction enzyme fingerprinting. *Genomics* 1992 **14:** 162-167
23. Tynan K, Olsen A, Trask B, De Jong P, Thompson J et al: Assembly and analysis of cosmid contigs in the CEA-gene family region of human chromosome 19. *Nucleic Acids Research* 1992 **20:** 1629-1636

6

Genome programmes and medical genetics: the case of the fragile X syndrome

Back to "reverse genetics"

This strategy whose proper – though less generally accepted – designation is "positional cloning" [3] follows the steps indicated in Figure 6-1. These can be initiated once the hereditary character of the disease has been ascertained, as indicated by the family tree at the top of the diagram. First, affected kinships are subjected to genetic analysis in order to search for linkage between the disease and polymorphic markers whose positions on the chromosomes are known. Numerous probes must be used at this phase (especially if there is no prior indication on the chromosome involved), combined with Southern blotting of DNA samples if RFLPs are employed, or with numerous PCR reactions in the case of microsatellites. This ultimately results in the "localization" of the disease, i.e. in finding convincing linkage between the disease and one or two polymorphic markers, that define an interval in which the gene involved must be located.

The second step, then, is a detailed study of this region, which spans at best several hundred kilobases, at worst several thousand. The mapping and cloning techniques described above are applied to define a detailed map of the region underscored by cosmid or YAC contigs. Genes in this neighbourhood can then be identified through the application of a series of criteria. A very useful one is the presence of CpG or HTF islands [11], that have high concentrations of CpG dinucleotide, contain most of the sites cut by rare site restriction enzymes and are generally associated with genes [14]. In addition, fragments whose sequence remains essentially unchanged between the human and the mouse are sought since

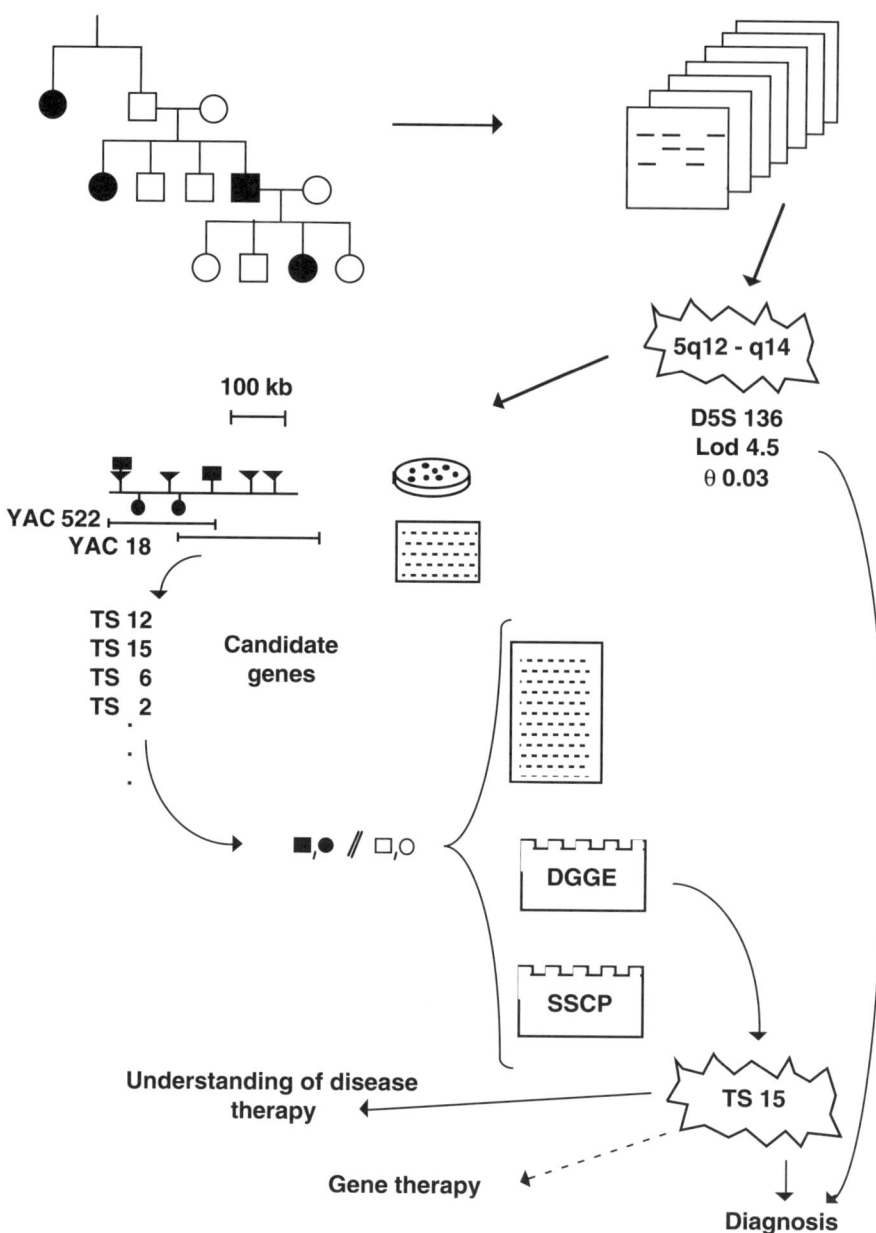

Figure 6-1 The steps of "reverse genetics" or "positional cloning" [3]. Research starts with affected families demonstrating the hereditary nature of the disease (top left). Genetic mapping (symbolized by a series of Southern blots) results in the "localization" of the disease gene but not (yet) its cloning. Physical studies (filters, Petri dishes) map out the region and detect several genes there (TS 2, 6, 15, etc.), which then become the candidate genes. By going back and comparing these genes in patients and controls (sequence, denaturation gradient gel electrophoresis, single-strand conformation polymorphism, etc.), the gene involved in the disease can be identified and subsequent steps undertaken.

they often correspond to expressed sequences. Hybridization of YACs or cosmids with messenger RNA extracted from different tissues can also be performed, not to mention more sophisticated methods such as "exon trapping" that uses biological selection to detect open reading frames [2]. This yields a series of candidate genes each of which, given its location in the area demarcated by genetic analysis, may be the sought-after gene.

Obviously the problem is then to determine which of the candidates is the correct gene: this third step entails going back to patients. Preceding studies could be conducted on "standard" DNA, obtained from some reference cell line since, apart from minor differences, the genome of all individuals is the same and the number, order and position of the genes are invariable. However, to determine which of the genes is responsible for the disease, these small variations now become of paramount importance because one of them is responsible for the ailment and hence will lead us to the gene ! Should one of these five or ten "candidate genes" be consistently missing (deleted), mutated or non-expressed in patients but present and active in their healthy parents, brothers, sisters, etc., then this will provide powerful evidence for its implication in the disease. Thus a comparative analysis is needed of the structure of the genes in patients and controls – bearing in mind that a minor variation may be the mutation responsible for the dysfunction or, on the contrary, a simple polymorphism without any functional impact.

One of the most notable successes of this method was the identification of the cystic fibrosis gene in a race that mobilized the resources of numerous laboratories. After detailed study of a 500 kilobase region, which genetic analysis predicted "should" contain the cystic fibrosis gene, Lap-Chee Tsui and his collaborators in Toronto [20] examined several candidate genes and retained one that featured a deletion of three nucleotides in patients – the well known delta F 508, resulting in the deletion of an amino acid (phenylalanine) in the corresponding protein. Furthermore this gene was expressed in the pulmonary epithelium, a critical organ in these patients, and its sequence appeared similar to a constituent of the ion channel implicated in the ailment – all of which constituted solid proof, and this identification has indeed resisted the challenge of time.

The contribution of Genome programmes

Despite its attractiveness, reverse genetics has drawbacks due to its narrow focus. The primary and almost unique goal in a programme like that of the Toronto team is to identify one gene. The chromosome area surveyed is relatively vast because the genetic localization is imprecise; its size can only gradually be reduced to focus attention on the most promising kilobases. In the months or years it takes to achieve this, a great deal of fragmentary data is obtained and numerous DNA clones are superficially examined; only those proving most pertinent to the stated objective are kept and studied in detail. The result is that many of the primary data are never studied in detail and remain fragmentary, since they are not germane to the objective. The same phenomenon occurs in the other laboratories that are simultaneously using similar methods to isolate the same gene. By adding the sums invested in all the laboratories involved, the total cost of cloning the cystic fibrosis gene has been estimated at $50-100 million – a sum that would certainly have been sufficient to fund a complete genetic and physical map of chromosome 7 (containing the CFTR gene) and to make gene isolation a straightforward operation. Instead, large areas of this chromosome remained virtually unexplored (though it is now being methodically mapped in the USA at Saint-Louis). Many more examples could be quoted and it is obvious that an approach focusing solely on genetic diseases tends to survey certain "hot points" in great detail while leaving uncharted vast tracts... that probably contain the hot points of the future !

A glance at Figure 6-1 illustrates how a rational build-up of genome data can benefit reverse genetics. For the localization of a disease, it goes without saying that the more complete the pre-established genetic map, the finer its grid and the more its markers are polymorphic (hence informative in the families of patients, often of small size), then the easier it will be to find convincing linkage. For the cloning of the corresponding gene, things are even more obvious: if the region indicated by genetic analysis is already available as a contig of YACs or cosmids, it is directly accessible. The candidate genes are already known or can be very quickly found by studying the cloned DNA segments. Only the third and last step remains to be performed, in which the patients' DNA must be analyzed to identify the disease gene.

The long march to the fragile site on chromosome X

An enigmatic syndrome

Interactions between medical genetics and the genome programmes will be illustrated by considering the fragile X-linked mental retardation syndrome [8, 9], a hereditary affliction that long defied the concerted endeavours of molecular biologists and clinicians. Indeed, the systematic correlation of this serious mental deficiency with a fragile site at the interface of the last two bands of the long arm of chromosome X was established a long time ago – the first paper demonstrating this association dates from 1969 [15], which means the syndrome was in fact already localized at that time. In comparison, this stage was only reached for cystic fibrosis in 1985 and merely four years were necessary to progress to the isolation of the gene in the autumn of 1989 [20].

The correlation between the mental retardation syndrome and the presence of the fragile site implied that the gene involved had to be located in, or very near to, the fragile site found in patients at the interface of the q27 and q28 bands of chromosome X. The obvious next step was to clone the DNA sequence corresponding to the fragile site and its vicinity, then to identify the genes present there and determine which one was absent, inactive or mutated in the patients. Yet cloning, and even approaching this fragile site proved very difficult: up to 1989 the nearest probes were still at a genetic distance of ten to fifteen centiMorgans, thus, presumably, separated from the goal by a comparable number of megabases of DNA, a huge distance even with modern DNA analysis methods [10]. In 1989 and 1990, several teams succeeded in isolating more strategically located DNA segments which were then employed to obtain the corresponding YACs. This step-by-step process ultimately yielded YACs that spanned the region of the fragile site and were found to contain unstable sequences in a gene called FMR-1 (Fragile-X Mental Retardation). The malfunction of this gene after amplification of the unstable sequences is in all likelihood the cause of the syndrome.

"Fishing for probes"

A look at this phase illustrates a common feature of research in medical genetics, the dearth of DNA segments located in a particular interval and the means used to remedy this. Though different tactics have been employed to isolate such probes, their common goal in this case was to obtain cloned segments coming from the narrowest possible area around the fragile site. The conventional method consists of isolating clones from a library constructed from DNA of a human/hamster or mouse hybrid containing part of the human X chromosome – in those days the best available was a cell line containing only the Xq24-Xq28 region. Clones are then tested on DNA from a series of cell lines containing various deleted or translocated X chromosome in order to determine their localization. This thankless and labour-intensive work produced new probes – the closest being located within a few centiMorgans of the fragile site.

Other teams selected the promising microdissection technique, which involves cutting out a small fragment of the X chromosome (containing the fragile site) and then obtaining cloned DNA segments from this material. As the amounts of DNA recovered are extremely small, PCR amplification is absolutely necessary but somewhat special insofar as the sequence to be amplified is not known. The choice of primers is therefore problematic, but this hurdle can be circumvented. Known sequences can be ligated at the ends of the dissected fragments, or the presence of repeated sequences can be relied on to ensure amplification in the presence of the corresponding primers. As described by Bernard Horsthemke's group [16], this method was applied to the fragile site by Kay Davies' team at Oxford and somewhat differently using laser microdissection by our own team in collaboration with a Japanese group [4]. In both these cases, clones were obtained and, though their analysis proved tricky because of their very short length, they did provide an approach to the fragile site.

In the course of 1990, these various strategies resulted in the isolation of numerous probes within an interval of five to ten centiMorgans around the fragile site. These landmarks then had to be positioned with respect to the site as well as to the sequence involved in the disease, which need not coincide exactly. Genetic analysis applied at this stage proved rather ineffective as distances were small and recombination frequencies very low; two loci cannot be ordered unless a recombination is found in the interval separating them in a family where they are informative. Thus the number of families needed to find such a recombination grows very quickly as the size of the interval is reduced, thereby making the use of other methods mandatory.

Steve Warren's translocations and the importance of "panels"

Cytogenetics has often come to the rescue of scientists studying genetic diseases. If one is lucky enough to find a patient whose ailment is associated with a deletion of a chromosome region or with a translocation involving two chromosomes, there is a good chance that these rearrangements affect the gene in question. By using the powerful existing techniques based on comparison of normal DNA with patient DNA, these rearrangements can be analyzed molecularly and the sought-after gene isolated quite directly. Thus, a small deletion observed on the short arm of chromosome X played a decisive role in the cloning of the dystrophin gene involved in Duchenne muscular dystrophy. In the case of the fragile-X syndrome, the fragile site itself gave a good indication of the probable position of the gene but no known translocation or deletion was available to facilitate access to the area - and it was therefore necessary, as it were, to generate them on purpose.

This was the aim of ingenious experiments conducted by Steve Warren (Figure 6-2) and his team for several years in Atlanta. Assuming quite reasonably that the fragile site should be a privileged breakpoint for chromosome X, Warren performed fusions of carefully chosen cells to produce a somatic hybrid containing a human X chromosome from a fragile X individual in a hamster cell. He could then grow this hybrid under conditions promoting chromosome rearrangements (e.g. in the presence of caffeine) and next, using previously introduced metabolic markers, select the cells in which a rearrangement had occurred in the fragile X region. He was able to produce a series of hybrids in which such translocations had occurred and that contained either an almost complete human X chromosome, connected to a hamster chromosome at Xq27.2 (the approximate position of the

Figure 6-2 Some of the main characters in the saga of fragile X. The top photograph shows Jean-Louis Mandel (left centre) and Steve Warren (right centre) at the Cold Spring Harbor Symposium in 1989. In the middle, two snapshots taken during the Oxford Fragile X meeting in September 1990 showing: on the left (left to right), Jean-Louis Mandel, David Schlessinger and Grant Sutherland; on the right Kay Davies (centre). The lower picture shows Grant Sutherland in Adelaide, Australia, June 1991. (All photos by the author.)

fragile site), or the opposite combination [24]. His idea was to employ these hybrids for direct cloning of the fragile site. In principle, all that was needed was to construct a DNA library from one of these cells, then to look for the clones containing a composite DNA segment, part human and part hamster. This clone would then correspond to the translocation point and, in principle, to the fragile site – an imaginative approach that was not to succeed as such, mainly for technical reasons. However, the resulting translocated chromosomes, in which the translocation point did in fact coincide with the fragile site or its immediate vicinity, provided a very effective marking tool for ordering the many probes described above.

Precise ordering of a series of DNA segments known to lie in an interval of a few megabases is not a trivial task: it is still very difficult to construct pulsed-field gel maps over such distances, and general YAC contig building over the human genome is a very recent development, not available at that time. However hybrid cells such as those described above can be used to form "panels" which are very fast and accurate to use. A single hybridization of the probe on a Southern blot containing the DNA of a translocated chromosome (e.g. Xpter-Xq27.2), previously isolated in a hamster cell, fixes the position of the probe above (positive hybridization) or below (negative hybridization) the translocation point. Despite some predictions that somatic hybrids would be supplanted as mapping tools with the advent of *in situ* hybridization for single copy probes, they actually proved highly effective for ordering a set of probes distributed throughout a small chromosomal region, provided of course a set of adequate hybrids was available. Those researchers who had not made a real effort to assemble such resources were subsequently to regret it... As for fragile X, thanks to the hybrids established by Steve Warren and several others, the laboratories that had set up good panels were able to position the several probes then available in the vicinity of fragile-X and to quickly select the most promising.

YACs come to the rescue

No expert insight was needed to predict that YACs were to play a major part in the final approach to the fragile site. Given the total lack of information regarding the type of gene to be identified – some scientists maintained that there was no protein-coding sequence involved, or even that the fragile site contained no DNA – the first priority was obviously placed on isolating a cloned DNA segment running from one "edge" of the site to the other. Opinions differed widely as to the size of the DNA sequence forming the site but the generally accepted figure was several hundred kilobases and the YACs seemed cut out for the job. The laboratories with well placed probes therefore made use of the existing libraries, notably at Saint-Louis, at CEPH and somewhat later at ICRF.

It is likely that the first YAC spanning the fragile site was isolated in a joint programme involving the teams of David Schlessinger in Saint-Louis, who with David Burke, Georges Carle and Maynard Olson set up the first human YAC library, and the Australian cytogeneticist Grant Sutherland (Adelaide), who had rediscovered the fragile site in 1976 (Figure 6-2). This clone was found to have a completely unexpected structure, i.e. circular and not linear as any self-respecting YAC ought to be ! This considerably slowed down its analysis until the fact was understood, but reports began to circulate in the summer of 1990 indicating that this YAC, used as a probe for *in situ* hybridization on chromosomes featuring the fragile site, did in fact label both sides of the site. This indicated that in all probability the YAC extended from one side to the other and that the size of the region did not exceed 200-300 kilobases. Thus isolation of the site appeared to be within reach, and a series of papers published in the first half of 1991 did indeed report the characteristics of the YACs containing the site. The first paper was mainly the work of Jean-Louis Mandel's team in Strasbourg [7] and it was followed a little later by the Australian team [25] and by a joint US/Dutch venture headed by David Nelson in Houston and Ben Oostra in Rotterdam [22]. In addition, the last paper reported the identification of the gene, FMR 1 for Fragile X mental retardation 1, suspected to be directly involved in this syndrome.

Hypermethylation and instability

However the YAC front line was not the only one to move forward as the data obtained before YAC isolation had already indicated that the fragile site was at long last within reach. The first sign was the detection of areas of hypermethylated DNA. For several years, many laboratories had been searching for indications of DNA hypermethylation in the fragile site region. A series of experimental indications and Charles Laird's very promising model (integrating the "imprinting" concept to explain the syndrome's unusual genetic features [13]) led to the belief that one or more genes in the vicinity of the site might be hypermethylated in affected individuals. This DNA modification, which generally correlates with the inactivation of the gene affected [11], was supposed to intervene in the mechanism of the disease. However, it could also be useful as an indication of proximity to the fragile site.

Until mid 1990, studies using pulsed field analysis of DNA digested by methylation-sensitive enzymes (only cleaving non-methylated DNA) had not produced conclusive results. These studies are rather tricky as the methylation state of a DNA sample depends on many factors, including for example the growth conditions of the cells from which it is isolated. In any case, the probes used at that time to detect the DNA fragments were more than one megabase away from the fragile site, hence outside any potential hypermethylated area. However, a paper published in mid-February 1991 by Jean-Louis Mandel's team demonstrated that one of the clones recently isolated by the group detected a clear and systematic difference on pulsed field gels between control and patient DNA, suggesting that the latter contained a hypermethylated area [23]. This paper was quickly followed by another reporting similar – but more detailed – results from the British team of Kay Davies, another specialist of the fragile X syndrome [1].

A look at the submission and publication dates of these two papers highlights the intensity of the competition in the "hot" fields of human genetics as well as the importance of maintaining good connections with the editors of scientific journals. The paper of the French team was submitted to *Nature* on October 25, 1990, accepted on December 7 and published on February 15, 1991 – nearly four months later. In contrast, the paper of the British laboratory was sent to *Cell*, which normally has similar processing deadlines. Initially received on January 15, 1991, a revised version submitted on February 4 was published on February 22nd... Priority, which may have a considerable media and commercial impact, is thus subject to manipulation; the Anglo-Saxon dominance of all major biological journals (except for *EMBO Journal*) can be definitely detrimental to "outsiders".

Fragile X begins to be understood

It soon became apparent that the fragile X region contains a particular DNA fragment that is longer in the patients than in the controls [17]. Grant Sutherland's team was able to correlate this instability with the variation of a short sequence consisting of the CCG trinucleotide (or CGG depending on the strand considered), repeated in tandem a few times in the DNA of healthy subjects but much longer (200-1,000 repeats) in the patients. This sequence was contained in the 5' region of FMR 1 gene described by Verkerk et al [22]. It could thus be expected that its elongation would interfere with the functioning of the FMR 1 gene, in accordance with a previously proposed scheme [9], and possibly cause the appearance of a fragile site as well as hypermethylation of adjacent DNA. The progressive expansion of this CCG stretch over two or three generations provided a plausible explanation for the strange inheritance pattern of this syndrome.

Numerous subsequent studies were to confirm these results and to clarify the nature – if not the mechanism – of the genetic defect. The FMR 1 gene is indeed central to the disease: it is not expressed in most patients [18], and a singular case of a person showing a typical Fragile X phenotype without the fragile site was resolved recently by the demonstration of a 2.5 megabase deletion in Xq27.2 that removes the gene [5]. Thus it is almost certain that the absence or inactivation of FMR 1 is the cause of the syndrome. Inactivation is caused by amplification, or lengthening, of the CCG sequence – through a yet unknown mechanism –, that also creates the fragile site and, directly or indirectly, causes hypermethylation of the region. The unusual genetics of the syndrome, with an increase in the number of affected individuals in successive generations, and the existence of normal transmitting males, are all connected with progressive amplification of the CCG repeat – although the exact relationship between the number of repeats and the phenotype remains to be determined. In practice this new knowledge has spawned DNA diagnostics, with or without the use of PCR, that directly determine the number of CCG repeats. This method is much faster and more reliable than cytogenetic examination or family studies. It also allows – at least in most cases [6] – the detection of carrier females and even of normal transmitting males, a considerable improvement. Thus, after a very long unrewarding period, molecular studies of Fragile X have finally paid off.

In addition, this research has for the first time demonstrated the existence of a mechanism involving not just a simple mutation or a deletion, but a variation in the length of a region of DNA, a mechanism which has since been implicated in Steinert muscular dystrophy [12], Kennedy disease and Huntington's chorea [21]. This new feature underlines the dynamic characteristics of the genome, too often seen as a very fixed entity [19].

Take-home lessons

A few comments are in order on the story of fragile X syndrome. First of all, its solution is clearly and primarily due to the resources implemented for the genome projects; this is particularly true for the YAC libraries which were set up to satisfy a global objective but proved extremely valuable in this case as in many others. Other technical fall-out from these programmes, such as sophisticated pulsed field gels capable of separating very large DNA segments and the general availability of probes and information, also played a more discreet but nevertheless real role.

Another conclusion is that coordinated projects – in principle based on wide-ranging collaboration – have had little impact on the very competitive mentality of geneticists. In fact, all the conditions were combined in this instance to promote a race between laboratories: an important syndrome, because of its high frequency of occurrence (roughly one boy in 2000), with therefore large impact in the media and easy availability of families, and a complex transmission mechanism indicating a new and exciting genetic mechanism, as subsequently confirmed. The fact that almost no headway had been made for several years was tempting everyone to try their hand as the leading teams in the field were in reality hardly further ahead than the others. We ourselves fell into this trap since we competed, more than we collaborated, with the main French group involved in this field, Jean-Louis Mandel's team, one of the three still there at the finish.

A certain degree of competition is certainly useful as it guarantees that results are checked. It also leads to the implementation of a range of strategies: if one of them turns out to be a dead end, progress will not be held up for long. However, above a certain threshold, this competition becomes excessive – apparently the case for the fragile X syndrome, which was tackled by dozens of teams, often as their main research project. A small one-day meeting on this topic at Oxford in September 1990 drew close to 50 participants representing over 30 laboratories – and several North American groups were not there ! One of the objectives of the genome programmes is to reduce the waste inherent in this type of unbridled competition; however, in practice this proves difficult because the customs, organization structures and recognition mechanisms prevailing in the community all tend to encourage competition.

REFERENCES

1. Bell MV, Hirst MC, Nakahori Y, Mackinnon RN, Roche A et al: Physical mapping across the fragile X: hypermethylation and clinical expression of the fragile X syndrome. *Cell* 1991 **64:** 861-866
2. Buckler AJ, Chang DD, Graw SL, Brook JD, Haber DA et al: Exon amplification: A strategy to isolate mammalian genes based on RNA splicing. *Proc Natl Acad Sci USA* 1991 **88:** 4005-4009
3. Collins FS: Positional cloning: let's not call it reverse anymore. *Nature Genetics* 1992 **1:** 3-6
4. Djabali M, Nguyen C, Biunno I, OostraBA, Mattei M-G et al: Laser microdissection of the fragile X region: identification of cosmid clones and of conserved sequences in this region. *Genomics* 1991 **10:** 1053-1060
5. Gedeon AK, Baker E, Robinson H, Partington MW, Gross B et al: Fragile X syndrome without CCG amplification has an FMR 1 deletion. *Nature Genetics* 1992 **1:** 341-344
6. Haberman R: Clinical conundrums in fragile X syndrome. *Nature Genetics* 1992 **1:** 157-158
7. Heitz D, Rousseau F, Devys D, Saccone E S, Abderrahim H et al: Isolation of sequences that span the fragile X and identification of a fragile X related CpG island. *Science* 1991 **251:** 1236-1239
8. Jordan B, Mattei JF: Retard mental lié à la fragilité du chromosome X: où en est-on en 1989? *Médecine-Sciences* 1989 **7:** 450-458
9. Jordan B: Fragile X mental retardation and the difficulties of reverse genetics. *Bioessays* 1991 **13:** 243-251
10. Jordan B: Megabase methods: a quantum jump in recombinant DNA techniques. *Bioessays* 1988 **8:** 140-145
11. Jordan B: Ilots HTF: le gène annoncé. *Médecine/Sciences* 1991 **7:** 153-160
12. Junien C, Lavedan C: Dystrophie musculaire de Steinert: encore une mutation instable. *Médecine/Sciences* 1992 **8:** 249-251
13. Laird CD: Proposed mechanism of inheritance and expression of the human fragile-X syndrome of mental retardation. *Genetics* 1987 **117:** 587-599
14. Larsen F, Gundersen G, Lopez R, Prydz H: CpG islands as gene markers in the human genome. *Genomics* 1992 **13:** 1095-1107
15. Lubs HA: A marker X chromosome. *Am J Hum Genet* 1969 **21:** 231-244

16. Ludecke HJ, Senger G, Claussen U, Horsthemke B: Cloning defined regions of the Human Genome by microdissection of banded chromosomes and enzymatic amplification. *Nature* 1989 **338**: 348-350
17. Oberle I, Rousseau U F, Heitz D, Kretz C, Devys D et al: Instability of a 550-base pair DNA segment and abnormal methylation in fragile X syndrome. *Science* 1991 **252**: 1097-1102
18. Pieretti M, Zhang F, Fu YH, Warren ST, Oostra BA et al: Absence of expression of the FMR 1 gene in fragile X syndrome. *Cell* 1991 **66**: 817-822
19. Richards RI, Sutherland GR: Heritable unstable DNA sequences. *Nature Genetics* 1992 **1**: 7-9
20. Riordan JR, Rommens JM, Kerem B, Alon N, Rozmahel R et al: Identification of the cystic fibrosis gene: cloning and characterization of complementary DNA. *Science* 1989 **245**: 1066-1073
21. The Huntington's disease collaborative research group: A novel gene containing a trinucleotide repeat that is expanded and unstable on Huntington's disease chromosomes. *Cell* 1993 **72**: 971-983
22. Verkerk A, Pieretti M, Sutcliffe J, Fu Y, Kuhl D et al: Identification of a gene (FMR-1) containing a CGG repeat coincident with a breakpoint cluster region exhibiting length variation in fragile X syndrome. *Cell* 1991 **65**: 905-914
23. Vincent A, Heitz D, Petit C, Kretz C, Oberle I, et al: Abnormal pattern detected in fragile X patients by pulsed-field gel electrophoresis. *Nature* 1991 **349**: 624-626
24. Warren ST, Zhang F, Licamelli GR, Peters J: The fragile X site in somatic cell hybrids: an approach for molecular cloning of the fragile site. *Science* 1987 **237**: 420-423.
25. Yu S, Pritchard M, Kremer E, Lynch M, Nancarrow J, Baker E, Holman K, Mulley JC et al: Fragile X genotype characterized by an unstable region of DNA. *Science* 1991 **252**: 1179-1181

7
Getting down to sequencing ?

Initially, the genome programme had been "sold" to the US Congress as a major sequencing project [21]. Deciphering our genotype was perceived to be a high-tech endeavour, like the Apollo Project, requiring comparable funding (several thousands of million dollars) and equally likely to confirm US supremacy in science and technology. At the time, it was anticipated that private companies would carry out the work, then sell the sequences to firms interested in using them. This aspect is once again in the news due to the current controversy over patents for cDNA sequences as well as the announcement of new commercial sequencing projects led by entrepreneurs such as Frederick Bourke [6] or others [7]. However, as most people are now aware, sequencing is but a small part of these programmes. Existing technology is not even close to providing a near-term solution for the complete determination of a sequence that is three thousand million nucleotides long.

Widespread optimism

Not long ago, in the late eighties, many scientists and policy makers were still very optimistic and predicted that large DNA regions would be sequenced in the very near future. The "sequencers" marketed by Applied Biosystems, later by Du Pont and Pharmacia/LKB were supposed to provide a quantum jump in productivity while making unnecessary the use of radioactive isotopes. Furthermore, robot systems were expected to quickly automate most of the associated procedures, and sophisticated data processing software would, it was believed, easily take care of the assembly and verification of sequences as well as their subsequent interpretation. Given this context, programmes aimed at sequencing genome regions spanning one or even several megabases, such as the major histocompatibility complex (two megabases), the q28 band of the X chromosome (ten megabases) or the complete T cell receptor locus in man and mouse were discussed, and sometimes launched, without excessive anguish.

A standard strategy

All the groups involved adopted a similar strategy [33] with some minor variations. The sequencing unit was a fragment of DNA cloned in a phage (15-20 kilobases) or cosmid (30-40 kilobases). This large DNA segment was subdivided into sub-clones in "shotgun" fashion, using random DNA breakage by sonication followed by sub-cloning of the fragements. Other, more systematic fragmentation methods were available but they were all considered too labour-intensive for large-scale use. The process continued with preparation of single-stranded templates, sequence reactions with fluorescent primers or nucleotide precursors and analysis of the products using an automatic sequencer (generally an "Applied Biosystems" machine). Capable of determining 5,000-10,000 "raw" nucleotides per day (24 tracks, 400 nucleotides read per track), each of these machines was supposed to produce over one million raw nucleotides per year. Since 5-10 kilobases of such data are usually required to assemble one kilobase of confirmed sequence, an output of several hundred kilobases per year (using sophisticated software for assembly of all these small sequences into a large one) was expected for laboratories using several of these machines.

Disillusionment sets in

Results after one or two years were, in fact, very much below the target: for a long time the largest published sequence remained that of the Epstein-Barr virus, 172,000 nucleotides hand-crafted in Great Britain in 1984 [9], followed in 1990 by the 229 kilobases of the cytomegalovirus (CMV), also sequenced manually [11]. Only around 1992, four or five years later, did such projects succeed in producing continuous sequences spanning a hundred kilobases or more. In fact, the scientists in charge of these operations had run into a multitude of technical and organizational problems arising from the magnitude of the attempted leap forward and from the difficulties inherent in partial automation.

First of all, the "sequencing machines" only actually automate one step of the sequence determination: the separation of the DNA fragments on gel and their detection. Moreover, they are very demanding in terms of the quantity and quality of the reaction mixtures. Whereas the manual method can manage with a fraction of a microgram of low purity DNA, the machine requires a larger amount of a virtually perfect sample – which may explain why some of them sit, imposing but idle, in quite a few laboratories! This requirement for quality DNA – a common feature of all models to a greater or lesser extent – is especially unwelcome here because it would be desirable to automate prior ("front-end") steps, such as DNA preparation from clones. But the preparation methods producing very pure DNA almost always include centrifugation, a manipulation that is difficult for robots to perform; while techniques involving filtration can be automated more easily, they usually yield DNA of lower purity. This is an example of a bottle-neck, a serious problem in such enterprises. A complex procedure cannot proceed more quickly than the slowest of its steps and virtually nothing is gained by accelerating one part of the process if others are still manual and slow. The experience of the megasequencing laboratories has been that as soon as one bottle-neck was eliminated, another one turned up, with the result that working on the project became like running an obstacle race.

Technology is not the only problem

Another complication of a different type arises from interference between "production" (obtaining the sequences) and "development" (improving the sequence determination process). These two functions are a potential source of hold-ups, particularly when performed by the same persons on the same machines. For instance a few tests per week can easily tie up half of the machines, thereby significantly reducing production. On the other hand, if development is neglected, the risk is that a great deal of time and money will be wasted because the techniques used are allowed to become obsolescent. Finding the ideal trade-off is no easy matter! Another organizational problem concerns the personnel: as this type of work is not really interesting in itself, who should carry it out? Two schools of thought exist on this subject: the first favours the selection of relatively unqualified technicians who will undergo in-house training and are expected to carry out these repetitive tasks over a prolonged period. The second aims for a higher level of competence and accepts from the start that the person hired is unlikely to stay for more than a year or two before moving on to a more challenging job. Each solution has its advantages and drawbacks; they will be discussed further in Chapter 13, centred on "sociological" issues.

A fresh start on a more reasonable basis

Megasequencing laboratories have learnt their lessons from the difficulties experienced and have lowered their sights; they have also recognized the need for an industrial-type organization, particularly well described in a recent paper from Leroy Hood's group (Figure 7-1) [36]. One of the important features of such a set-up is the

Figure 7-1 Two important personalities in DNA sequencing. In the upper photograph Leroy (Lee) Hood, one of the biologists most involved in the development of new instruments and responsible (with others) for major sequencing projects. In the lower photograph, George Church, inventor of the multiplex method. (Photos by the author.)

separation of production and development through the establishment of a distinct development team in which researchers and engineers can experiment with methods without day-to-day interference with production. Greater attention is also being paid to machine maintenance and repair using local resources without having to call in manufacturer representatives. Detailed studies are conducted on all phases of the process to obtain realistic estimates of the output, to optimize the overall system and to focus effort on the most limiting steps. All the above considerations would not be out of place in an automobile plant: this remark underscores the qualitative change needed to switch to this kind of scale. In other words, it is not simply a question of quantitative scaling-up but, instead, the whole process has to be organized in a different way.

An ambitious but realistic sequencing project: the Nematode

The Nematode *(Caenorhabditis elegans)* sequencing project, addressing an organism whose 100 megabase genome is about the same size as an average human chromosome, serves as an excellent test for the strategies and methods and provides a good example of a second-generation programme [35]. Conducted jointly by a British team (Alan Coulson and John Sulston at Cambridge) and by a US group (Bob Waterston at Saint-Louis), the programme has the short-term goal of determining a total sequence span of 3 megabases in the centre of chromosome 3 (a region rich in genes) and the long-term objective of sequencing the entire genome of this model organism. A strict timetable calls for each team to cover 100 kilobases in the first year, 400 kilobases in the second and one megabase in the third. Noteworthy features are the planned increase in output as well as the tactics employed, i.e. a combination of random and directed sequencing using a total of four sequencers at the Cambridge end of the collaboration.

In outline, each cosmid is fragmented into several hundred sub-clones in phage m13, the single-stranded DNAs are prepared in microtiter plates in four or five days. And then the reactions are analyzed in a week on the two Applied Biosystems sequencers. This provides a three- to four-fold redundancy, leaving "gaps" in the sequence (points without any sequence) and "problems" (areas with poor quality or contradictory data). These uncertainties are then resolved by directed sequencing involving the synthesis of oligonucleotide primers from the previously determined sequences: they are dubbed "walking primers" since they make it possible to move from known to unknown sequences. Sequencing at this stage is performed with the Pharmacia machines, better suited for this phase since they only use one kind of fluorescent label, making custom primer synthesis relatively easy. Roughly one month is needed to perform this process for the hundred or so problem spots and to finish the sequence of one cosmid. During that period the next cosmid is being sub-cloned and analyzed on the Applied Biosystems machines. A total of seven or eight persons are needed, including two for assembling the sequences, using locally developed software which, during checks of machine-performed alignments, is capable of re-displaying the original data from the machines on the same screen within two seconds.

On the whole, this programme is an excellent example of megasequencing in the nineties, seriously and resolutely conducted, irreproachably organized while not involving a surfeit of technology or equipment. Indeed, initial results [35] from this programme have turned out to be interesting and quite surprising: many previously unknown and unsuspected genes were found in this organism, though both it and its genetic map were thought to be well known.

Yeast chromosome III: a successful consortium

Yeast has been graced with the honour of being the first to have one of its chromosomes completely sequenced. The project was organized in a rather unusual way. A European consortium of 35 laboratories was formed, and the sequence was shared out in twelve-kilobase chunks, each group receiving two Ecus (2.5 US dollars) per confirmed base determined. At the outset, early in 1989, scepticism was rampant: the programme drew criticism from some quarters as lacking in ambition, since chromosome III spans barely 315 kilobases, less than twice the 172 kilobase Epstein-Barr virus genome sequenced by hand in 1984 [9] by a single British laboratory. Simultaneously, it was expected that coordination of such a large number of teams with quite different levels of expertise would be extremely difficult, and that the project would fail or at least run late. In any case, while I was travelling in the USA during the first half of 1991, I did hear much about the still tentative US plans for yeast genome sequencing. But I do not recall anyone – not even David Botstein, one of the most ardent defenders of this project – mentioning the European enterprise which at that time was already well underway.

As it turned out, the groups in the European consortium did fulfil their quota, the coordinating centre in Martinsried (Germany) did assemble the sequences, and by mid-1991 work was essentially complete. A total of 385 kilobases of finished sequences had been obtained for this 315 kilobase chromosome, almost all of it by manual methods. Comparison of the 35 kilobases sequenced twice by different laboratories gave an error rate of 0.4‰, i.e. one every 2,000 bases, quite a reasonable figure. The sequence was submitted in December 1991 and published in May 1992 [30]. Again, as in the case of the Nematode, many "new" genes were revealed: close to 150, on a very well-studied chromosome where a total of fifty were known and a few additional ones expected. Since introns are almost absent in yeast, recognition of genes from the DNA sequence is straightforward. The figure obtained, a gene every two kilobases, is a very solid one – all the more since it has been possible to show that almost all of them are actually transcribed. Some of these "new" genes show sequence similarity with sequences already present in the databases and found in man, *Drosophila*, *Xenopus* or even *Arabidopsis*. Thus the "minimum set" of coding sequences required for such a simple organism as *Saccharomyces cerevisiae* increases, and it is now expected that a total of six to seven thousand genes may exist in this "lower" eucaryote. All told, the 2.65 million Ecus (approximately 3 million US dollars) spent for this programme (including clone procurement, informatics and coordination of the work) have not been wasted, and the results are well worth this amount. It is however obvious that megasequencing will have to move away from consortiums of a very large number of small groups if costs are to be brought down to a more reasonable level. This is in fact planned for the next European step aimed at chromosomes II and XI, representing a total of over one and a half megabases. Meanwhile, chromosome I is being tackled in Canada, V in the USA and VI in Japan: complete sequencing of the yeast genome is in sight.

Craig Venter's interests extend beyond cDNA

Craig Venter's laboratory in Bethesda is mostly known nowadays for massive partial sequencing of cDNA clones... and for a very debatable decision to attempt patenting these sequences. However this very team had planned sequencing the whole q28 band of the X chromosome – i.e. approximately ten megabases – back in 1988. This project was not funded, but it is not surprising to find the group reporting results of genomic sequencing. A 1992 paper [27] reports a 106 kilobase sequence determined around the ERCC1 DNA repair gene in chromosome 19, the chromosome chosen by Anthony Carrano's laboratory at Lawrence Livermore for cosmid contig building. Three of these cosmids have been sequenced by Venter's group, using shotgun subcloning and Applied Biosystems sequencers.

Assembly of the sequences was made difficult by the presence of numerous repeated elements, and additional fine restriction mapping experiments were necessary to resolve ambiguities. Finding the genes was not trivial: this is human DNA, and the genes are made up of many small exons scattered over large tracts of DNA. According to the author's own estimates, the computer programs used had a false negative rate of 60% – i.e. they missed more than one exon out of two – but they did usually find several exons for each gene. And the other pieces could then be tracked down by closer examination. Five genes were ultimately found in this 116 kilobase region: the previously known ERCC1 gene, a fosB proto-oncogene, another whose product resembles a phosphatase, and two new genes that show no similarity to known sequences. The harvest is definitely less bountiful than with yeast or nematode DNA sequencing, it does however point to a total of 75,000 to 150,000 genes in the human genome. Similar conclusions were reached in other recent mammalian genome sequencing exercises conducted by Lee Hood's [36] or, again, Craig Venter's [28] groups. These results show that it is now definitely possible to sequence hundreds and possibly thousands of kilobases in man, but that the task is both more arduous and less rewarding than for simple organisms whose genome is much more compact.

Genome sequencing comes of age

It has thus become practical to sequence a megabase of human DNA. This was believed to be the case four or five years ago: at that time, it was wrong, but advances in technology and foremost in organization have now changed the picture. A budget of one megadollar for such an endeavour is not unreasonable – providing the segment in question is already entirely cloned in cosmids. Current software is able to detect a sizeable proportion of exons, although defining the limits of genes in human DNA remains tricky. The estimated total number of genes keeps increasing. Each completely sequenced region has revealed more, often many more genes than expected. In experimental organisms, most of them seem to have so far escaped detection by genetic analysis because their inactivation has no apparent consequence on the phenotype – at least under laboratory conditions. This does not mean, however, that they are useless... What about man? To what extent can we still believe the current estimate of 50,000 to 100,000 genes when we discover that a nematode, with its 959 cells and 302 neurons, has 15,000 genes ?

From a more "utilitarian" point, it becomes tenable to think about genome sequencing as a way of finding genes that have been located by genetic mapping. At this stage, researchers usually face a region spanning one or a few centiMorgans, a few megabases, within which the "morbid" gene must lie. Several methods can be used to catalogue the "candidate genes" existing in this interval, but each of them suffers from specific drawbacks and tends to miss many genes. Why not sequence the whole region and obtain thus a catalogue of the exons it contains ? Such an approach has already been used, e.g. by Christine Petit's group (with vital help from Généthon) at the Pasteur Institute to find the Kallmann syndrome gene [26]. It is true that the interval had been narrowed down to less than 100 kilobases, but it is not absurd to imagine a tenfold increase of this figure.

cDNA studies are very popular

An effective approach

Sequencing cDNA appears at first sight to be a very reasonable approach, especially for studies on man. Such an approach focuses on the most interesting areas, those corresponding to the genes, by-passing the vast stretches of neighbouring "junk DNA" that have no known function except possibly a mechanical role in conserving chromosome structure. The technical means for direct access to gene sequences is provided by cDNA libraries, constructed using messenger RNA extracted from tissues or cell cultures and then copied *in vitro* into DNA. Conventional genetic engineering techniques can then be used to propagate the DNA in bacteria. By definition, this cDNA contains sequences corresponding to genes; it is therefore a favoured tool for acquiring gene data. However the various genes are non-uniformly represented in cDNA libraries due to the highly variable amount of the various species of messenger RNAs in the cell, which results in a high degree of redundancy for the corresponding clones in the cDNA library. This difficulty, believed some time ago to be a very serious one, appears to have been solved; several laboratories have demonstrated that these libraries can be "equalized" or "normalized" so that the various sequences are more or less uniformly represented [25, 31]. In addition, the abundance of individual messenger RNA species in itself can constitute useful information, so that normalization is not necessarily advantageous. The cDNA approach was one of the points stressed in the British, Japanese and French Genome Programmes, but US teams have been quick to adopt this line of attack with their usual efficiency.

The sequence as a "signature"

What appears currently to be the most productive procedure is to pick clones randomly from a cDNA library, and to determine a small amount of sequence for each: 200-300 nucleotides, i.e. a sequencing "run" on an automatic sequencer, that analyzes several dozen samples in parallel (Figure 7-2). This information is incomplete since the full sequence of a cDNA usually comprises 1,000-10,000 nucleotides. Furthermore, it is somewhat approximate, with a possible error rate of 1-2%. The partial sequence, often called a "signature", is nevertheless extremely valuable as it can be compared with sequences already recorded in the databases to ascertain whether it matches a known gene. It can also be used to derive part of the structure of the corresponding protein; it makes possible the localization (or at least the chromosome assignment) of the gene. Last of all, the partial sequence allows, if necessary, the isolation of the complete gene as a cosmid or a YAC clone, with a substantial amount of additional work, but with virtually certain success.

This strategy is extremely effective: a well organized, medium-sized laboratory, with a staff of a dozen and a yearly operating budget of $1 or 2 million, can expect to process several thousand cDNA clones per year. As, in 1992, only approximately 3,000 of our estimated 100,000 genes were known, it is clear that this approach has a very significant contribution to make. If this line is pursued (as is presently the case) by several teams, it will quickly provide at least partial information on most

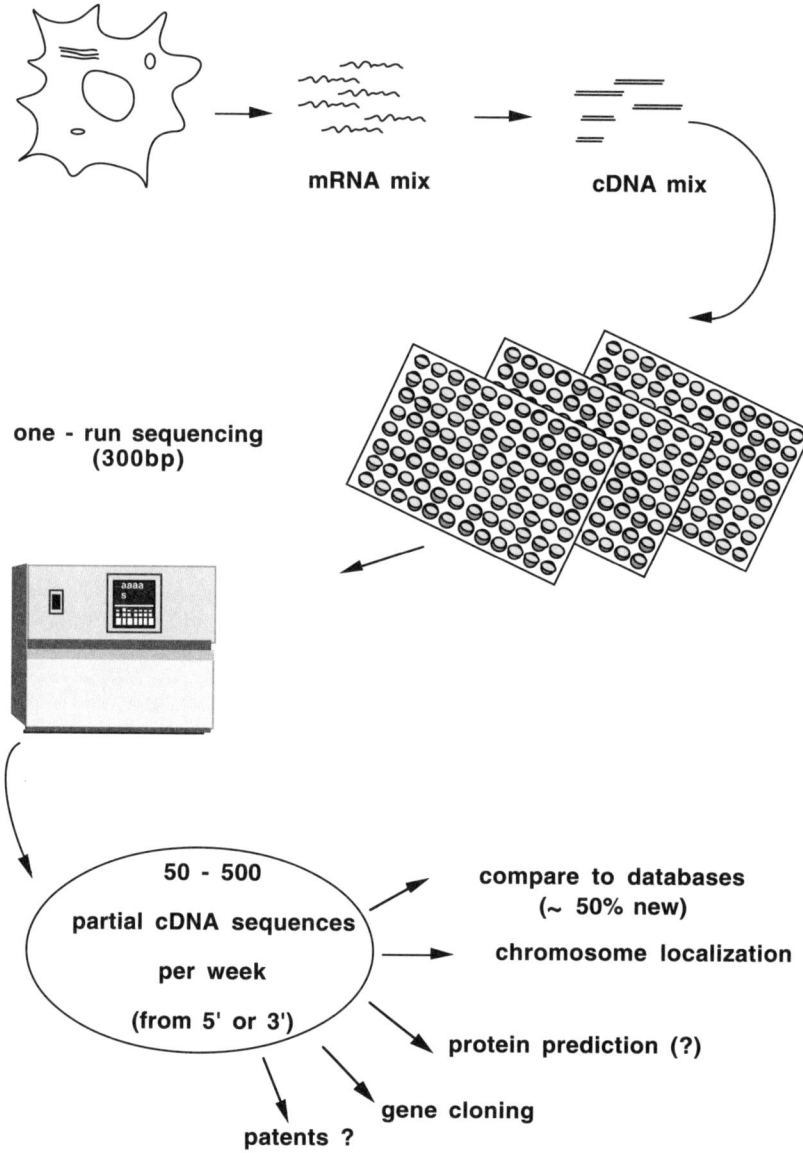

Figure 7-2 Massive partial sequencing of cDNA clones. The messenger RNA mixture extracted from a cell is copied into DNA and cloned. The bacterial colonies are arranged on microtiter plates. Randomly selected clones are then subjected to a sequencing "run", i.e. one migration in a sequencer which can read 300-400 bases with an error rate of 1-2%. The machines and processes currently in use have high throughputs. The partial sequences are next compared with those already entered in the data bases. Other possible operations on these sequences are chromosomal localization (though this is much more time-consuming than the sequence), partial prediction of the sequence of the corresponding protein and, if appropriate, isolation of the gene itself.

of the genes in our genotype. Craig Venter's group, which was responsible for popularizing this "signature" method [1] that had already been proposed several years before by Sydney Brenner in the United Kingdom, later provided an additional 2,375 brain sequences [2]; meanwhile another US group [23] determined a further 1,024 and mapped a sizeable proportion of them; Japanese teams, notably Kenichi Matsubara's in Osaka [29], have set out on a plan to determine the "cDNA profile" for the two hundred cell types that constitute our body by sequencing 1,000 cDNA clones from each of them. Other publications, from Charles Auffray's "Genexpress" programme at Généthon or from the British Resource Centre are expected; most of the corresponding sequences are already in the public databases. A rough estimation, late 1992, indicated that already close to 20,000 "signatures" had been obtained by the various groups pursuing this approach all over the world - interestingly, when these teams compare their data they find so far very little overlap, so that these 20,000 signatures probably represent an almost equivalent number of different genes.

Enemy brothers ?

Are genomic and cDNA sequencing opposed or complementary ? cDNA sequencing "à la Venter" provides quite cheaply many small pieces of information, which may build up rapidly a catalogue of all known and unknown genes transcribed in the tissue from which the cDNA library was made. However this efficient approach provides neither the complete sequence (necessary for structure and - *a fortiori* - function predictions) nor the position of the gene, and it may give a very sketchy view of the genome. Genomic sequencing, on the contrary, does provide detailed and complete data - at a price. A price that, in present-day conditions, is too high to compete with cDNA sequencing for our own large and complex genome. On the other hand, for yeast and even the nematode, the genomic approach is probably the most profitable. A modest improvement in present technology, such as consistent reading of 1,000 or 2,000 bases per sample instead of the present 400, could make a lot of difference...

A fight over patents

No discussion on this subject can be complete without mentioning patents or, more exactly, patent applications, as it is far from certain that they will be granted. Craig Venter's employer, the National Institutes of Health (NIH), or more precisely Reid Adler as director of the NIH office of Technology Transfer, registered a patent application for first a few hundred and later more than two thousand new sequences obtained. The aim was protection of the interests of (North-American) firms wishing in the future to develop products based on the genes to which these sequences correspond. Will patents actually be granted ? Nothing is less certain:

to be patentable, an invention must be an *innovation*, it must not be *self-evident* in terms of the current state of the art and its *utility* must be fully apparent. How novel is a sequence that already existed in the genome before it was deciphered? It is *discovered* rather than invented and a discovery cannot in principle be patented. The second requirement for patentability is that the method should be original and not obvious to a specialist in the field, which once more does not seem to be the case. As for the utility of the invention, it can only be stated in very vague terms for the very compelling reason that the sequence is far too partial to even think about predicting the exact function of the corresponding gene and protein. Of course one must be cautious of the differences between ordinary (or scientific) language and legal concepts, and it is true that DNA sequences have already been patented in the USA and possibly elsewhere – but these were usually complete sequences, incorporated in a process expected to yield a marketable product. In addition the operation only makes sense if patents can be taken on batches of hundreds or thousands of sequences, otherwise the exercise becomes prohibitively expensive if each of them has to be individually patented.

Many scientists in the USA and elsewhere, including Jim Watson, then NIH Genome Program Director, strongly opposed this move which, they believed, would considerably slow down the dissemination of results and bring about enormous waste. The outcome of this controversy, whose pros and cons continue to be hotly debated [3, 20, 24] is still not clear. Various statements by governments and organization such as HUGO have strengthened the hand of those opposing "commercialisation" of the human genome. Nonetheless, the British Medical Research Council gave notification in early 1992 of its intention to file patent applications for tactical reasons, though in principle it was against this course of action, as explained by the Minister of Research [22]. As for Jim Watson, he resigned in late April of the same year from his position as Director of the NIH Genome Programme, apparently to a large extent because of his stand on this question. The US patent office has turned down the application, but this is only the first round, NIH may appeal the decision and it is quite common for patents to be awarded after initial rejection: the issue is still very much open.

Exotic techniques

In spite of many small improvements, the present DNA sequencing methods clearly do not measure up to the task of deciphering the human genome, i.e. three thousand million nucleotides (a figure to be contrasted with the ten or so million nucleotides of human sequence known at this time). All-out automation of the existing process has been attempted with some success, notably in Japan, and we will discuss the

"HUGA sequencing factory" in Chapter 9, which deals with instrumentation. Most of the present research effort, however, focuses on perfecting methods and, if possible, replacing them by faster and more cost-effective techniques based on another principle. "Evolutionary" developments attempt to improving existing processes, e.g. by electrophoresis of DNA in capillaries or ultrathin gels (10-100 microns) designed to enhance both speed and resolution. However, more exotic approaches are also being explored, such as attempts to read gels at ultra high speeds by laser-induced ablation of DNA labelled with stable isotopes (B. Jacobson at Oak Ridge), or the process developed by R. Keller (Los Alamos) in which a single DNA molecule is suspended in a liquid jet and an ultrasensitive system detects in real time the individual detached bases as they pass by (Figure 7-3). Yet other methods are aimed at extracting sequence data through

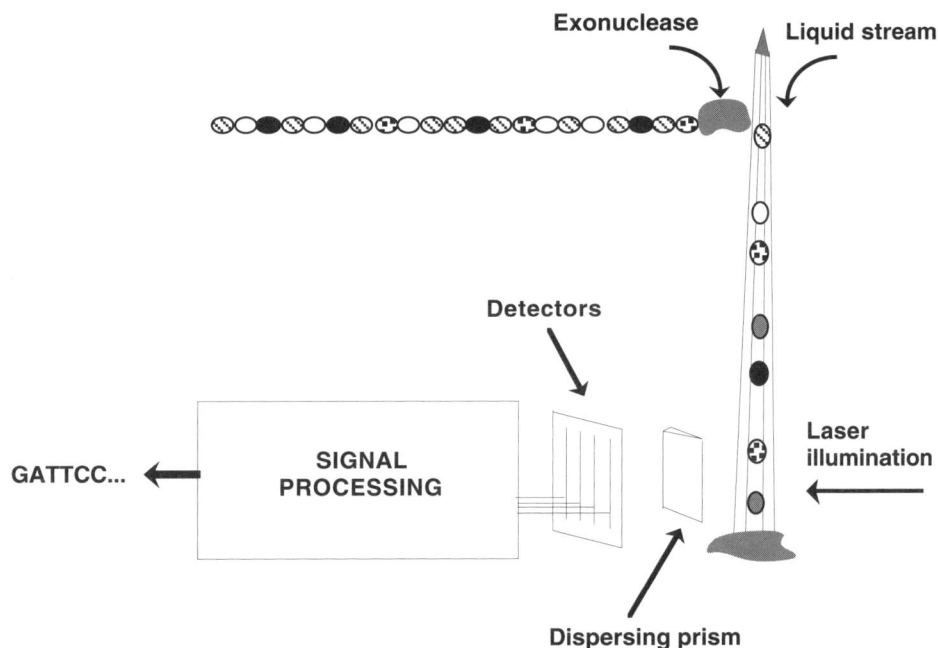

Figure 7-3 DNA sequencing by ultra high sensitivity detection of individual bases. The DNA molecule at the top is attacked by an exonuclease which successively detaches the bases, that have previously been labelled by conjugating a different kind of fluorescent group for each of them. The individual bases are then carried along inside a series of microdrops (as in a cell sorter) and registered as they pass by the detector. The fundamental feature of this method is that one single DNA molecule is analyzed, thereby eliminating any enzyme synchronism problems but imposing an extremely high performance detection system, capable of sensing the fluorescence generated by a single fluorophore molecule. Such performance is theoretically feasible and appears attainable in practice. The potential interest of this method is that it could be very fast with an output of several thousands of bases per second. Gibco BRL is involved in the development of this technique. (Diagram based on an illustration from Richard Keller's group at Los Alamos.)

numerous hybridizations with defined sequence oligonucleotides [10, 13, 32, 34] – a method which could be applied in a massively parallel fashion thanks to the use of techniques derived from microelectronics. We will discuss in some detail a process that has raised high hopes and earned wide publicity: scanning tunnelling microscopy. In theory, this method should be able to "view" the DNA molecule in enough detail to recognize its bases and hence its sequence; but its practical implementation is still beset with problems.

Tunnelling microscopy: a hope for DNA sequencing ?

Scanning tunnelling microscopy and related techniques

Scanning tunnelling microscopy became very fashionable for DNA analysis in the late eighties: every genome programme devoted a paragraph to it and quoted at least one laboratory working on its application to DNA sequencing. It is true that potentially very important applications exist in biology for the Scanning Tunnelling Microscope (STM) and related instruments, the Atomic Force Microscope (AFM) and the Scanning Ion-Conductance Microscope (SICM) – which explains the prevailing enthusiasm. In these instruments, a very fine tip is kept extremely close to the object under study by a control system based either on the detection of a minute current induced by tunnel effect between the tip and the object (STM) or on the measurement of Van der Waals interaction force between the probe and sample (AFM). The tip is then moved to scan across the object, its vertical displacements being precisely measured and stored, thereby generating a picture of the sample surface. In this context, the "very fine" tip is a mere one atom wide at the end, the "minute" current is measured in pico or nanoamperes and the tip-to-sample distance is of the order of one atomic diameter. Amazing as it seems, it really is possible to view in this way atoms, molecules and their arrangement on the substrate, whose perfect flatness is one of the method's critical points. Moreover the STMs or AFMs are not, as might be expected, enormous and astronomically priced machines but sell for fifty thousand US dollars or thereabouts, often less than an ultracentrifuge !

DNA sequencing by scanning tunnelling microscopy ?

It is not difficult to imagine what this technology could contribute to biology, particularly because the STM and AFM (originally restricted to an ultra-vacuum) can now operate in air and even in a liquid. There are many possible applications, including DNA sequencing, since the four bases have different chemical structures that a method with atomic scale resolution should be able to differentiate. Examination of DNA with the STM or AFM has been a very popular exercise, and has already been reported in numerous papers. Indeed, the DNA molecule is very attractive for near-field microscopy studies: it is prestigious, easily available and very resistant to degradation, more so than collagen or a protein. In addition, it is relatively easy to analyze. The unprocessed output of these methods consists of a series of profiles defining the vertical displacement of the tip; they are not always very illuminating, and this image must be analyzed in detail using powerful data processing equipment in order to see the object under study and to isolate its features from those of the support. Under these conditions, it is easier to detect the signal generated by a thread (the DNA molecule) running across the image than that due to a more or less globular object such as a protein molecule.

Let us indulge in daydreams for a moment and assume that a number of technical problems, discussed below, have been overcome. A sample carrying spread-out DNA molecules a few tenths of a millimetre long, YAC clones for example, (1,000 kilobases of DNA measure about 0.5 mm) would be placed on the sample stage of a "tunnelling sequencer". The machine would then start, find a DNA molecule and track along it, identifying bases on the fly, and generate a continuous 100-200 kilobase sequence in a matter of minutes or at the most hours. Science-fiction ? Certainly, but it is worthwhile to see how close present methods are to this goal.

High quality DNA imaging

The story of scanning tunnelling microscopy studies on DNA began in 1988 with the first paper from Travaglini's group [4], which showed good quality images of DNA bound to recA protein. The sample was coated with a very thin film of metal to provide the conductivity necessary, in principle, to induce the tunnel effect. The same team was to demonstrate a little later [5] that "uncoated" recA-DNA complexes could also be viewed. Even though the mechanism producing the tunnel effect is, in that case, not yet well understood, the result is there, and it opens up exciting prospects. Another encouraging aspect was that this latter work had been carried out in air rather than in vacuum. However, these results were obtained with DNA artificially "thickened" by complexing with protein, and the observation of native DNA molecules had not yet been demonstrated. This step was to be achieved in the same year with the publication of a paper by the US Department of Energy Livermore and Berkeley Laboratories, that included images of native DNA showing the large and small grooves of the double helix [12]. Even sharper images were published shortly afterwards by an Italian group [15]. In the same year a Minneapolis team [8] showed that scanning tunnelling microscopy could view the Z conformation of DNA in air and on naked DNA.

One big question is how routinely such results can be obtained – near-field microscopy of DNA has acquired a reputation for being extremely irreproducible. During my survey trip I had the opportunity of meeting some of these rather unusual microscope specialists. As head of the "Near Field studies group" at Lawrence Livermore, Rod Balhorn is a biologist with very good knowledge of DNA. Several persons, including physicists, are involved in this work which originally made use of a laboratory-built machine but which now employs a commercial device, the Nanoscope II from Digital Instruments. The group has extensively studied the adhesion of DNA to the substrate as this is one of the critical factors of these methods – in fact the tip tends to "pick up" the molecules during its scans. The group has also worked on imaging of the bases and produced good quality pictures of Adenine (monomers) organized in a 2-D crystal structure laid flat on the substrate. In another laboratory at Lawrence Berkeley, a team is studying the same topic in the Surface Research Department which has six scanning tunnelling microscopes. However, here again the lack of adhesion of the DNA to the substrate has caused problems and most of the team's current efforts are addressing this difficulty. At a DOE Contractor Meeting, the overall impression was that the subject was more or less marking time. After the initial outburst of enthusiasm spurred by the theoretical prospects of the method, everyone is now sadly conscious of the numerous artifacts found on the graphite support (the most commonly used), notably the "steps" that look very much like DNA molecules [14].

What remains to be accomplished ?

Assuming that these adhesion problems can be overcome, what remains to be done before the method can be used to sequence DNA ? Presumably it would be necessary to view single-stranded DNA, whose bases are exposed and more readily recognizable than in the double helix where they are hidden inside the structure – a justification for the publication on the front page of *Nature* in late 1989 [19] of the first credible picture of single-stranded DNA (poly (dA)). This raised hopes that it would at least be possible to differentiate between purines and pyrimidines, provided a solution was found for the difficulties experienced with molecule spreading and orientation. Since then the rush of papers has slackened and only one top grade image from Caltech (US) was honoured on the cover page of *Nature* [18]. These were very precise views taken of double-stranded DNA in an ultra-vacuum that not only show the large and small grooves but details of the structure as well (Figure 7-4). As in most of the preceding publications, this paper also concludes with a discussion on the prospects for sequencing DNA.

The major aim of work now being carried out by several laboratories is to actually identify bases; there is reasonable hope that this may be achieved in the near future with STM, or with AFM. Indeed, while most of the publications quoted above

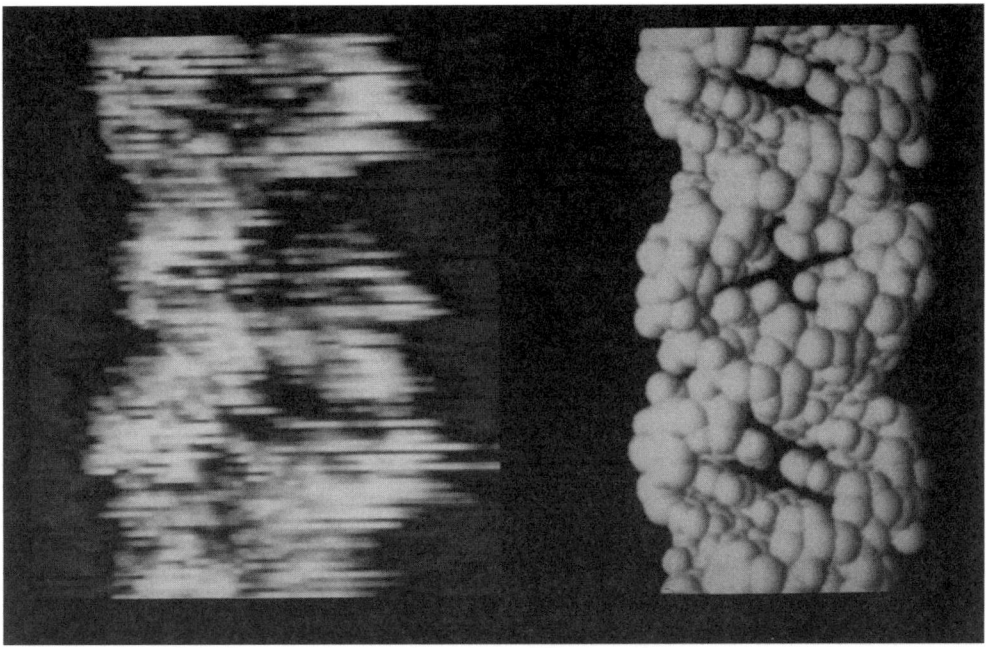

Figure 7-4 Scanning tunnelling microscopy image of DNA. Kindly made available by Dr John Baldeschwieler [18], this figure shows on the left an image of uncoated DNA taken by STM under a vacuum, and on the right a photograph of an atomic model of the molecule. The level of detail visible and the agreement with the model are both remarkable.

refer to scanning tunnelling microscopy, the more recent AFM features some impressive advantages, including elimination of the need for an electrically conducting sample. Technically these methods are very similar in terms of instrumentation, tips, mechanics, electronics and imaging analysis. The scanning ion-conductance microscope (SICM) and even a new type of optical microscope are all members of the same family called "near field microscopes"; in fact these analogous techniques are complementary. Everyone is now waiting for the first credible demonstration of base recognition on a sample, in all likelihood on single-stranded DNA and perhaps with chemically modified bases to render them more readily distinguishable. Interesting for both scientific and technological reasons, this objective is almost certainly attainable and is being relentlessly pursued by a host of teams.

Sequences, DNA-binding factors and interactions

The next step is much more nebulous. Assuming – as is plausible – that the bases can be recognized using STM, AFM or any other near field method, will this develop into a competitive sequencing technique? Reliability, speed and cost will then be of paramount importance: no longer will it be acceptable to spend hours computing on a powerful graphics workstation to enhance each image. Indeed, it seems very likely that the near field microscopy techniques will ultimately have the greatest impact in fields such as the binding topology of various factors (transcription, regulation, etc.) to DNA, or the dynamic study of protein-protein or DNA-protein interactions. It should not be forgotten that these methods – after all quite gentle – are suitable for examination of living matter and its transformations, as Drake et al demonstrated [17] by observing the polymerization of fibrin using AFM (see also on this topic an excellent "News and Views" by R. Crowther in *Nature* [16]). In any event it seems clear that a new era is opening up for microscopy, that calls not only for physicists and biologists to collaborate closely, but also for biologists to keep track of developments in this field.

"Small" laboratories should not be overlooked

In addition to sequencing specialists, there are many teams who need now and then to determine a DNA sequence spanning several kilobases or several dozen kilobases. This is a need which is also widely experienced outside the human genetics community and concerns all molecular biologists to a greater or lesser extent. Up till now, the vast majority of these teams continue to use conventional sequencing methods since the currently available machines are too expensive, too productive and also too disruptive as they make necessary many changes in the upstream and downstream procedures. However, considerable opportunities for innovation exist in this field as well as a large potential market, that interests manufacturers. Compared to the several dozen or so large megasequencing centres, there are literally thousands of laboratories performing small-scale sequencing and interested in making this easier, faster and less expensive. Manufacturers have of course noticed this opening, and Hitachi in Japan is reported to be on the verge of commercializing such a machine at about a quarter of the cost of the present ones. Other advances are taking place, such as the direct detection of radioactivity or the replacement of X-ray films by imaging plates. These plates are exposed (like a film) for up to several hours and then "developed" in a few minutes by a special device which reads them with a laser beam. Not only is this system quicker but the image is directly recorded in computer memory. Another option is to replace radioactive isotopes by photoluminescent labelling systems that could finally gain a place in DNA research laboratories. As for capillary electrophoresis, which has been a roaring success in biochemical laboratories, it could serve as the basis for sequencing units with lower output but a more affordable price than existing machines.

REFERENCES

1. Adams MD, Kelley JM, Gocayne JD, Dubnick M, Polymeropoulos MH et al: Complementary DNA sequencing: expressed sequence tags and Human Genome project. *Science* 1991 **252:** 1651-1656
2. Adams MD, Dubnick M, Kerlavage A, Moreno R, Kelley J et al: Sequence identification of 2,375 human brain genes. *Nature* 1992 **355:** 632-634
3. Adler RG: Genome research: fulfilling the public's expectations for knowledge and commercialization. *Science* 1992 **257:** 908-914
4. Amrein M, Stasiak A, Gross H, Stoll E, Travaglini G: Scanning tunnelling microscopy of recA-DNA complexes coated with a conducting film. *Science* 1988 **240:** 514-516
5. Amrein M, Durr R, Stasiak A, Gross H, Travaglini G: Scanning tunneling microscopy of uncoated recA-DNA complexes. *Science* 1989 **243:** 1708-1711
6. Anderson C, Aldhous P: Genome Project faces commercialization test. *Nature* 1992 **355:** 483-484
7. Anderson C: Controversial NIH genome researcher leaves for new $70-million institute. *Nature* 1992 **358:** 95
8. Arscott PG, Lee G, Bloomfield VA, Evans DF: Scanning tunnelling microscopy of Z-DNA. *Nature* 1989 **339:** 484-486
9. Baer R, Bankier AT, Biggin MD, Deininger PL, Farrell PJ et al: DNA sequence and expression of the B95-8 Epstein-Barr virus genome. *Nature* 1984 **310:** 207-211
10. Bains W: Hybridization methods for DNA sequencing. *Genomics* 1991 **11:** 294-301
11. Barrell B: DNA sequencing: present limitations and prospects for the future. *FASEB Journal* 1990 **5:** 40-45
12. Beebe TP, Wilson TE, Ogletree DF, Katz JE, Balhorn R et al: Direct observation of native DNA structures with the scanning tunneling microscope. *Science* 1989 **243:** 370-372
13. Cantor CR, Mirzabekov A, Southern E: Report on the sequencing by hybridization workshop (special feature, meeting report). *Genomics* 1992 **13:** 1378-1383
14. Clemmer CR, Beebe TP: Graphite: a mimic for DNA and other biomolecules in scanning tunneling microscope studies. *Science* 1991 **251:** 640-642
15. Cricenti A, Selci A, Felici AC, Generosi R, Gori E et al: Molecular structure of DNA by scanning tunneling microscopy. *Science* 1989 **245:** 1126-1127
16. Crowther RA: Probing biological structure. *Nature* 1989 **339:** 426-427
17. Drake B, Prater CB, Weisenhorn AL, Gould SA, Albretch TR et al: Imaging crystals, polymers and processes in water with the atomic force microscope. *Science* 1989 **243:** 1586-1589
18. Driscoll RJ, Youngquist MG, Baldeschwieler JD: Atomic-scale imaging of DNA using scanning tunneling microscopy. *Nature* 1990 **346:** 294-296
19. Dunlapp DD, Bustamante C: Images of single–stranded nucleic acids by scanning tunnelling microscopy. *Nature* 1989 **342:** 204-206
20. Eisenberg RS: Genes, patents, and product development. *Science* 1992 **257:** 903-908
21. Gall JG: Human Genome sequencing. *Science* 1986 **233:** 367-368
22. Howath A: Patenting complementary DNA. *Science* 1992 **256:** 11
23. Khan AS, Wilcox AS, Polymeropoulos MH, Hopkins JA, Stevens TJ et al: Single pass sequencing and physical and genetic mapping of human brain cDNAS. *Nature Genetics* 1992 **2:** 180-185
24. Kiley TD: Patents on random complementary DNA fragments? *Science* 1992 **257:** 915-918
25. Ko MSH: An "equalized cDNA library" by the reassociation of short double-stranded cDNAs. *Nucleic Acids Research* 1990 **18:** 5705-5711
26. Legouis R, Hardelin J-P, Levilliers J, Claverie JM, compain S et al: The candidate gene for the X-linked Kallmann syndrome encodes a protein related to adhesion molecules. *Cell* 1991 **67:** 423-435

27. Martin Gallardo A, McCombie WR, Gocayne JD, Fitzgerald MG, Wallage S et al: Automated DNA sequencing and analysis of 106 kilobases from human chromosome 19q13.3. *Nature Genetics* 1992, **1:** 34-39
28. McCombie WR, Martin-Gallardo A, Gocayne JD, Fitzgerald M, Dubnick M: Expressed genes, Alu repeats and polymorphisms in cosmids sequenced from chromosome 4p16.3. *Nature Genetics* 1992 **1:** 348-353
29. Okubo K, Hori N, Matoba R, Niiyama T, Fukushima A et al: Large scale cDNA sequencing for analysis of quantitative and qualitative aspects of gene expression. *Nature Genetics* 1992 **2:** 173-179
30. Oliver SG, Van der Aart QJM, Agostoni-Carbone ML, Aigle M, Alberghina L et al: The complete DNA sequence of yeast chromosome III. *Nature* 1992, **357:** 38-46
31. Patanjali SR, Parimoo S, Weissman SM: Construction of a uniform-abundance (normalized) cDNA library. *Proc Natl Acad Sci* USA 1991 **88:** 1943-1947
32. Pevzner PA, Lysov YP, Khrapko KR, Belyavsky AV, Florentiev VL, Mirzabekov AD: Improved chips for sequencing by hybridization. *J Biomol Struct Dyn* 1991 **9:** 399
33. Smith LM, Hood LE: Mapping and sequencing the Human Genome: how to proceed. *Biotechnology* 1987 **5:** 933-939
34. Southern EM, Maskos U, Elder JK: Analyzing and comparing nucleic acid sequences by hybridization to arrays of oligonucleotides: evaluation using experimental models. *Genomics* 1992 **13:** 1008-1017
35. Sulston, Du Z, Thomas K, Wilson R, Hillier L et al: The C. elegans genome sequencing project: a beginning. *Nature* 1992 **356:** 37-41
36. Wilson RK, Koop BF, Chen C, Halloran N, Sciammis R et al: Nucleotide sequence analysis of 95 kb near the 3' end of the murine T cell receptor alpha/delta chain locus: Strategy and methodology. *Genomics* 1992 **13:** 1198-1208

8

Genome and informatics: the infernal twins

Throughout my study, in the USA and elsewhere in the world, one theme kept continually recurring: the importance of data processing. This topic is not just given lip service: genome laboratories often have a sizeable informatics team with five to ten data processing specialists. Major data centres, such as the Genome Data Base in Baltimore, have several dozen staff members and are funded accordingly. Informatics, in fact, is now involved at all stages of genome research, from the collection of data to its storage and dissemination.

Multiple requirements: data entry and interpretation...

The methods applied to study the genome are increasingly making use of computers to acquire and process information. Examples abound: confocal microscopy, analysis of fluorescent *in situ* signals by a charge-coupled device (CCD) (which efficiently converts the image into an electrical signal), complex-cycle control of pulsed field devices, or even exposure of radioactive filters onto imaging plates which are later scanned by a laser beam to reveal the latent image. It then becomes natural to employ computers to further manipulate this information. In commonly used procedures, an enlarged photograph of a gel stained by ethidium bromide is superimposed manually with an autoradiogram of the same gel after transfer and hybridization, to find out which of the DNA bands have hybridized to the probe. Today the recorded images of these two experimental results can be called up from optical or magnetic disk memory and aligned electronically. It is then possible to correct image distortion, to adjust scaling factors and to apply objective band coincidence criteria while directly archiving the results. This is just one of many instances...

Computerized laboratory notebooks...

Both improvements to techniques and their growing automation amplify, as they are indeed intended to do, the volume of information, clones and data obtained: it becomes unfeasible to register and archive them by hand. Manual methods are only capable of handling information about a limited number of objects, especially in laboratories with varying research topics, in a fluid and very busy environment. Many group leaders have experienced the nerve-racking search for a clone isolated a year or two ago by a since-departed post-doctoral associate, an experience which is not even over once the clone has been found since the restriction map and "bit of sequence" produced by a student to define PCR primers still have to be located... If the number of DNA segments isolated and analyzed is increased tenfold or hundredfold without substantially improving their storage and the archiving of their characteristics, a potential catastrophe is in the making. There is thus a pressing need for a computerized laboratory notebook allowing – and preferably compelling ! – each team member to record results as soon as they are obtained, according to a carefully defined procedure. The objective is to be able, at any time, to call up all existing information about every previously studied object.

An example: the Lawrence Livermore Genome Center

I was able to see an advanced model of these new electronic laboratory notebooks at Lawrence Livermore, where Tony Carrano directs research on chromosome 19. As indicated in Chapter 5, the analysis consists of studying a library of cosmids isolated from this chromosome and attempting to assemble them into "contigs" (a series of clones featuring overlaps). This is achieved by direct acquisition of the "signature" of each cosmid (a set of restriction fragments) on a system derived from the Applied Biosystems DNA sequencer. By pairwise comparison of the signatures (roughly 8,000 to date), the system then deduces which cosmids are liable to form a contig. The latter is then catalogued in a computer system and can be displayed with all its characteristics on the screen – including its degree of reliability, defined by the amount of overlap observed and colour-coded on the screen accordingly. The contigs are next checked by *in situ* hybridization of the two end cosmids, which should appear to coincide on the chromosome given the resolution of this technique. Last of all the contigs are linked with each other using a YAC library. All this information, plus the "history" of each contig, is stored and can be accessed; it is thus possible to look up result trends, to apply changeable but consistent criteria and to perfect the strategy adopted. Developed by Elbert Branscomb's group, this remarkable system is an example of positive interaction between informatics and biology. The group in which this work was carried out belongs to DOE, and this is probably not irrelevant: the high technological level and relative stability of the staff prevailing in this laboratory have helped to achieve this too rarely observed symbiosis.

Semi-private or semi-public databases...

However, results and clones are not for the exclusive use of the laboratory that produced them: they must be disseminated to outside collaborators and made available (hopefully as soon as possible) to the scientific community in accordance with a precise procedure and after a stringently defined validation process. This raises the question of data bases, their architecture and their relations; their organization must be highly flexible to cope with the variety of "entities" to be classified – yesterday plasmids and cosmids, today YACs or STS and tomorrow perhaps P1 clones or sister chromatid exchange breakpoints. The connections between the different kinds of data are all-important: this explains the widespread use of "relational" data bases, such as the Sybase system, which are well suited to manage this kind of situation. In a relational data base, an item of information (e.g. the name of a clone) is only recorded once in a specialized "table". The main task of the software is to handle the connections ("relations") between a number of different tables in a very powerful and flexible manner, linking the name of the clone with its sequence, defining its position on a restriction map or providing details on the researcher who isolated it. New relations can be introduced at any time. Should the name of the clone change, then a single modification in the table containing it will suffice to update it in all "views" presented to the user by the software.

The larger teams are developing their own systems, generally incorporating one or several workstations, although a few diehards still remain loyal to their Hypercard software run on a Macintosh computer. These new setups – at least the most advanced – manage several subsets of data: those specific to the laboratory, still preliminary or too "hot" to be made available to potential competitors; those coming from or accessible by outside collaborators; and finally those whose access is unrestricted and that will shortly be transferred into public data bases. It is no easy matter to administer these different confidentiality levels. Inter-bank exchanges (and their related checks) create in fact sensitive problems in the fields of data processing, organization and even politics.

General-purpose banks...

These are meant to provide a collection of commonly accepted data, useful as a basis for subsequent work. This was the task, it will be recalled, of the Human Gene Mapping Workshops which, starting in the early seventies, were held every second year. During these meetings new results for each chromosome were presented and discussed; the various maps were updated, often in the midst of lively debate. Each of these "conclaves" produced a compendium (whose size inevitably grew with time) containing all the agreed-upon information about probes, polymorphisms, genetic and physical maps. This "bible" then acted as the reference until the next meeting was held. The ever-increasing volume of data brought quick recognition of the need to employ computers. Today this function is principally carried out by the Genome Data Base (GDB), managed by Peter Pearson and funded initially by the Howard Hughes Medical Institute and later by the National Institutes of Health and Department of Energy. Except for the DNA sequences, GDB contains all genome-related data ranging from probes to fragile sites and including maps, mouse data, the atlas of human genetic diseases produced by Victor McKusick, plus the fax numbers and names of persons to contact for clone requests (Figure 8-1). Well structured and with ample resources, this bank is very complete and extremely useful although not exceedingly user-friendly. Read-only access is open to anyone (providing the necessary hardware, software and high speed links are available) but data insertion is strictly controlled by a complex process in which the "chromosome editors" nominated in the Human Gene Mapping Workshops play a central role. Secondary "nodes" will make connection to GDB easier while reducing communication costs: one such node is already installed in the United Kingdom at Northwick Park in London and another at Heidelberg in Germany. Further nodes are planned in Japan and at Uppsala (Sweden).

... and political stakes !

GDB is not however unique in the world: Jean Frezal in France created the "Genatlas" system for the HGM 9 meeting (held in Paris in 1987), and other databases are operating elsewhere, e.g. in Japan. Some competition is clearly beneficial; moreover, it seems legitimate not to leave the monopoly of storing genetic data to an organization located in a country which already largely dominates this research sector, especially as some of its scientific leaders have demonstrated imperialistic tendencies. In a closely related field, the existence of two major DNA sequence libraries (the European EMBL and the US GenBank) has helped to stimulate a competitive spirit and to promote healthy pluralism. At this time the USA alone provide all the funding for the GDB, whose resources, a staff of about thirty and an annual budget of several million dollars, largely overshadow those of its "competitors". Furthermore the direct link with the Human Gene Mapping Workshops, for which GDB is the official library, forms as it were a "closed shop". Unsurprisingly, most of the chromosome editors who control data entry are Anglo-Saxon.

Yet there is room for others, particularly if a certain degree of complementarity is sought by setting up systems that specialize, for example, in graphic representation

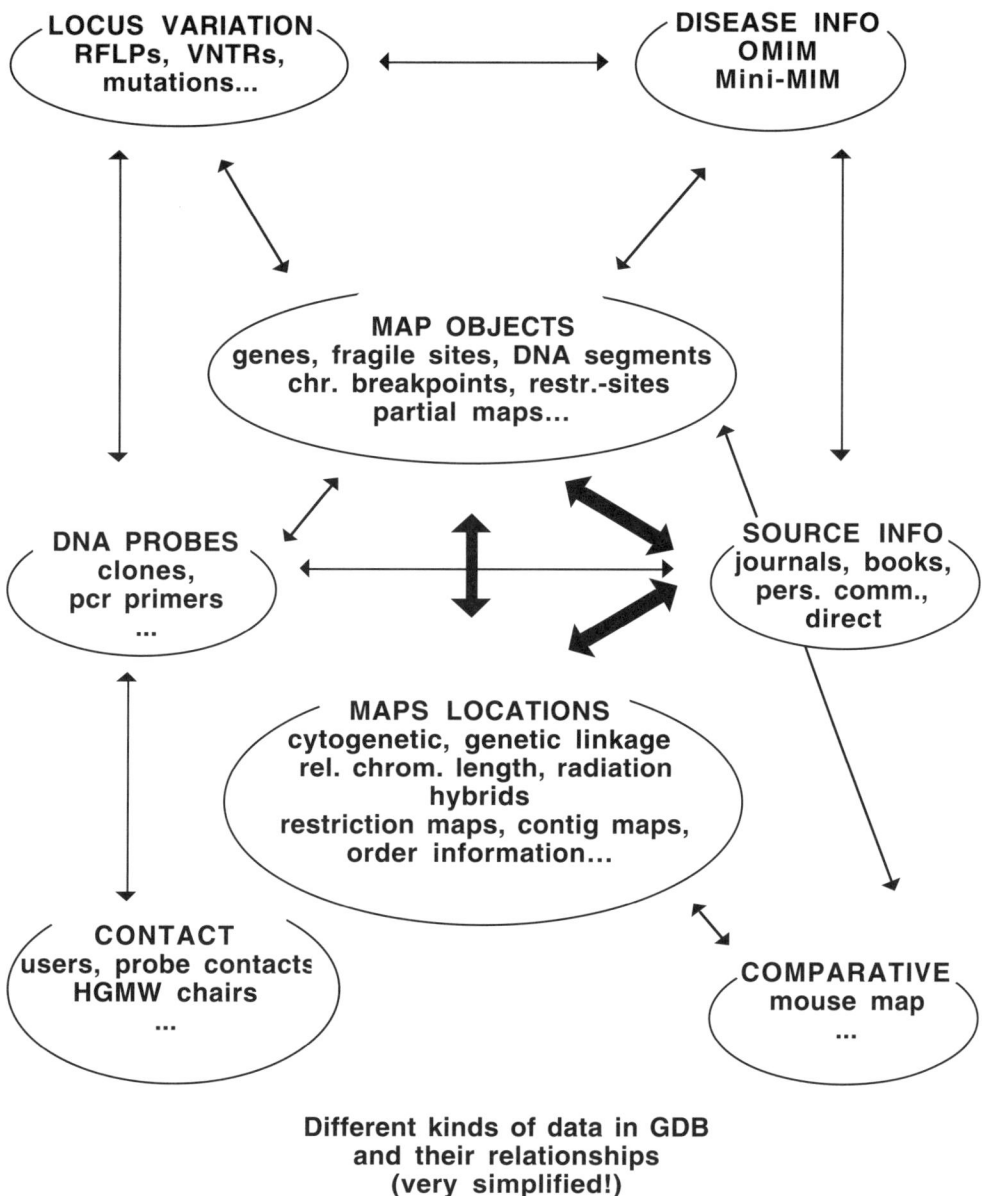

Figure 8-1 Structure of data in the Genome Data Base. Adapted from a publication by Peter Pearson, this diagram outlines the various types of data stored in the Genome Data Base and indicates the paths that can be followed from one kind of data to another. For instance, one can enter the library by specifying a chromosomal region, request which genes and anonymous segments known therein (top centre), consult their linkage maps (bottom centre), interrogate OMIM to ascertain whether genetic diseases have been located there and, lastly, obtain the names of the probes (left middle) and the address of the person keeping them (bottom middle).

of maps, in recording preliminary, unconfirmed results (nonetheless extremely useful and worth looking up), or that connect the contents of several different databases. Interestingly, a database originally developed for the Nematode community, ACeDB (A *Caenorhabditis elegans* Data Base) is gaining quite wide acceptance, has been adopted by the *Arabidopsis thaliana* genome project (under the name "AAtDB"), and even serves as the "front end" of an integrated genome database being developed in Heidelberg (Germany). The whole sector is currently in a continual flux with the increasing power of computing systems and the evolving requirements of users. As for GDB, its internationalization is now on the agenda and negotiations are underway with a view to joint funding (and management) by the USA, the EEC and Japan. It is vital for scientific communities outside the USA to have access to all the existing data under satisfactory conditions. Unfortunately Europe (and probably the rest of the world) has fallen considerably behind the USA in the field of high speed data links – a vital prerequisite for remote look-up of this kind.

Bicultural coexistence is difficult

The last question in our discussion is "extracurricular" to science but nonetheless highly important: how well do biologists and computer specialists work together ? Not exceedingly well, it seems. Even in the USA where substantial means are allocated for joint projects, this collaboration is not always trouble-free; one laboratory (whose name I will censor) had a superb database, used however solely by the computer specialist and the "boss", but completely ostracized by the post-docs, preoccupied as they were with their own project and little interested in overall group strategy ! To effectively integrate data processing calls for a meeting-up or, better still, a merging of two communities with very dissimilar educations, jargons and what can be called "cultures". Reasons for potential misunderstandings abound, emphasizing the need for a real effort to communicate. On one hand, biologists must strive to conceptualize and explain their needs, to attain a minimum level of computer literacy and to accept that a development period is mandatory for any fairly complex software. On the other hand, computer specialists must give priority to operationally rapid and effective solutions even when they are not optimum. They must be ready on occasion to use "quick and dirty" methods rather than aiming for sophistication and style in their software. Needless to say, this symbiosis is not found everywhere and, surprisingly, whether the head of a data processing group was originally trained as a biologist or a computer person does not seem to make a great difference – at least in the cases I have encountered.

106 TRAVELLING AROUND THE HUMAN GENOME

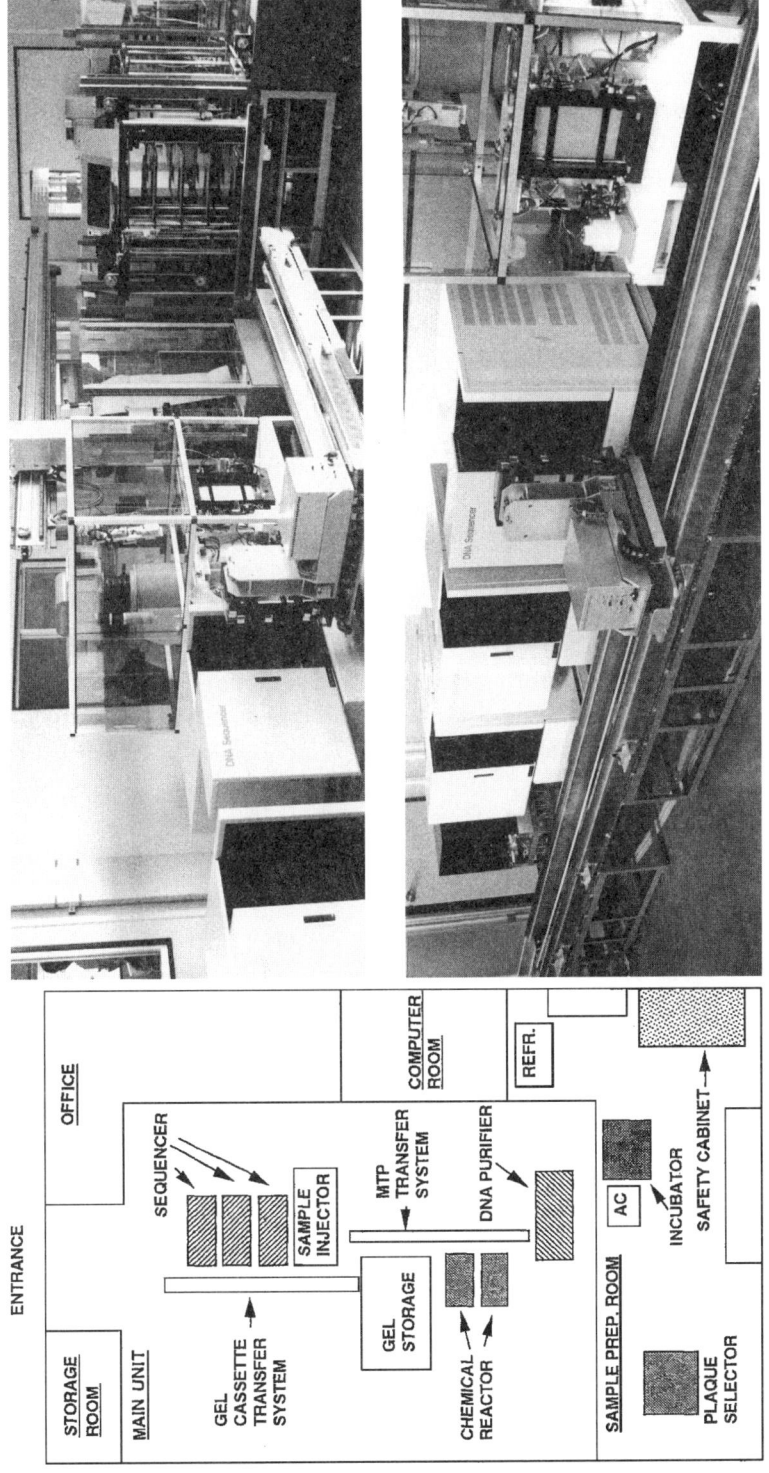

Figure 9.1 The HUGA "sequencing factory". The left-hand diagram shows the layout of the equipment in this assembly line, which handles all the operations from colony identification and sampling up to the recording of sequence data. The photographs on the right show two views of the installation: at the upper right, the gel pouring system, the gel loading robot, one of the three Hitachi sequencers and, in the foreground, the rail guiding the acrylamide gel cassette carrier; at the bottom, the end of the assembly line, showing the loading robot from another angle, the three sequencers, the transfer system and, at far left, the robot which separates the used gels from the plates onto which they were poured.

successively developed in Japan to automate this method: the first in 1983 and the second in 1986 by Seiko Instruments. Apparently, these machines did not find a market. Price was certainly a factor: the design was based on a liquid phase version of the method – very laborious to automate as several centrifugations are required – and not on its solid support alternative, more modern and certainly easier to perform in an instrument. Consequently the equipment was very complex and its cost, $300,000 in 1986, was really prohibitive.

Robotization of recombinant DNA technology is thus a high-risk endeavour: it entails the prospect of ending up with a machine either overtaken by technical advances or simply too costly for biological laboratories, since these usually consider a machine priced at $100,000 as a major expenditure: the same scale does not apply to our physicist friends ! Furthermore, it seems imperative not to just robotize the existing manual operations but instead to review the entire process, in order to find out whether it cannot be machine-performed in a different way – just like potato chip manufacturers who skin their tubers using high-velocity steam blasts rather than a vegetable peeler controlled by a computer coupled to an image analyzer. All this implies close partnership between industry and research, up-to-date technological surveys, motivated and efficient contacts and companies with sound financial backing.

Nevertheless some biological laboratories have succeeded in implanting robotics massively in their operations. I will discuss two of these cases that involve organizations operating in an unusual contex, which is probably not irrelevant to their success.

The HUGA "sequencing factory" in Tsukuba, ...

Without any doubt one of the most spectacular systems I saw during my survey was the "sequencing factory" built at the Riken Life Sciences Center at Tsukuba near Tokyo by a team under the leadership of Isai Endo and Eichi Soeda. This vast and lavish laboratory of the Science and Technology Agency (a government agency akin to the DOE in the USA or to the French CEA) has been for some time involved – among other things – in the development and testing of machines for DNA sequencing. In a previous visit, in late 1988, I had seen a Hitachi prototype sequencer rather similar to the Pharmacia/LKB machine, a robot for sequencing reactions and other pieces of equipment, all of which had little obvious superiority over models then marketed in the US. Today, however, a building in the Riken Center houses an astonishing installation: the very first automatic DNA sequence assembly line [3]. Occupying a room of about 100 square meters, the system (Figure 9-1) consists of a string of automatic machines: a robot to analyze the Petri

Automation, an obvious solution ?

In this age of microprocessors and robots, it appears deceptively simple to replace the often over-qualified staff performing these tasks (it is not unusual for a senior scientist to spend a morning pipetting microlitres of reagents into an array of tubes) by precise, rigourous, dependable and indefatigable automated systems. The fact that this robotization has hardly got underway illustrates that it is not straightforward – and there are both real and irrational reasons for this situation.

Let us start with the irrational ones. Surprising as it may seem, the biological research community is in general quite conservative: its members do not like to change methods unless forced to do so by the competition. How else can one explain, for example, the very slow penetration of non-isotopic DNA labelling techniques (the so-called "cold probes"), which feature many advantages and perform well in the vast majority of cases ? Each and every one has long been accustomed to using radioactive phosphorus and resists changing well-established habits. In fact, even when they communicate, researchers remain traditional; it is somewhat paradoxical that, in leading edge research at the fringe of the unknown, communication still consists mainly of the conventional typed text, which is meticulously set, printed and bound as in the time of Gutenberg: this is just beginning to change with the first electronic biology journal, the *Online Journal of Clinical Trials*. Another bad reason is that most biologists lack proficiency in technology and physics: replacing a fuse, using a multimeter or applying Ohm's law do not come easily to many of them. Hence communication with a robotics engineer is likely to be difficult. In France the absence of suitable middlemen does not help: almost all laboratory instruments are manufactured abroad (USA, UK, Sweden, Germany, etc.). Finding a dependable French industrial partner really interested in (and capable of) developing a product in this field borders on the impossible.

There are nonetheless justifiable reasons for this lag, the main one being that recombinant DNA techniques are still evolving very rapidly. The time required to develop and then commercialize a specialized robot, typically several years, means that it may be obsolete by the time it is marketed. A case in point: the methods based on enzymatic amplification (PCR) by-pass a series of steps and completely alter the level of sensitivity required in many applications since the region to be studied can be amplified specifically a million times beforehand. Thus Southern blotting is no longer necessary in many clinical diagnostic applications, and the market for an automatic blotting machine is much smaller than might have been predicted in the early or mid- eighties. Another instance, in the field of DNA sequencing, is the "Maxam and Gilbert" chemical degradation method [7], historically very important but now dethroned by the enzymatic techniques derived by Coulson and Sanger [8]. Two robots were

9
The end of "cottage industry" instrumentation ?

Research and manual labour

One of the first surprises experienced by the novice visiting a recombinant DNA laboratory is the widespread use of manual techniques. Some machines will indeed be found: ultracentrifuges, relatively sophisticated pulsed field gel elctrophoresis systems, PCR "machines"... In a very well off institution, a DNA sequencer might even be sighted – though often standing idle. Notwithstanding, the observer will be struck by the throng of persons performing what appear to be relatively low tech tasks: one individual is carefully capping small plastic tubes before installing them on a support cobbled together from old scraps of polystyrene and floating this contraption in a water bath; another one is meanwhile carefully shaking something that looks like a sheet of white paper in a Tupperware container filled with a hot, foamy liquid; while a third worker sits in front of a light box, pouring over an X-ray film covered with minute dashes and writing down arcane hieroglyphs on a sheet of paper. The reader may have recognized the set-up of an incubation, Southern blot washes and the analysis of a sequencing gel autoradiogram. In virtually all laboratories these tasks are indeed performed manually and machines – served by man – are only used from time to time for very specific operations. This manual work is repetitive, sometimes mind-destroying and not without mistakes. It is also critical because inattention or non-observance of the very stringent conditions at certain moments can invalidate the whole experiment and produce non-interpretable results.

dishes and pick the clones; a machine to make DNA preparations (in series of 24) from 2 ml cultures; two Seiko instruments (this model is now commercially available) to carry out the sequencing reactions; a machine to pour the gels; a robot to load the samples onto the gels and to initiate migration (and so save time on the sequencing run); three Hitachi sequencers (also now commercially available, at least in Japan); and, last of all, a machine to remove and dispose of the gels ! Except for the colony picker, all this equipment is integrated in the assembly line, which includes devices to handle the transfer of samples from one machine to the following one. The entire process is controlled by a computer workstation.

This "sequencing factory" is not fully automatic and involves a number of manual operations and checks which, according to Eichi Soeda, will require two or three persons full-time. A full cycle lasts for three hours (the various steps obviously overlap) and the design output is 20,000 nucleotides of raw sequence per cycle (16 samples, 450 nucleotides read, 3 machines), i.e. a production of over 100,000 nucleotides in 24 hours. When DNA sequencing is discussed, all theoretical production figures must be considered with a great deal of scepticism because they have always proved far too optimistic in the past [5]. But this time the system is very different, almost fully integrated and with allowance made for bottle-necks (the reason for the two reaction robots and the three sequencers); thus its real output may come close to the theoretical target. A trial run devoted to complete sequencing of the lambda phage (48,000 nucleotides), in order to assess the error rate, was originally planned for late 1991 or early 1992. In fact teething troubles have hit this set-up – not a completely unexpected occurrence – and it was not yet really operational in early 1993. There is still hope, however, and the sequencing factory may indeed end up churning out millions of nucleotides of sequence...

According to Eichi Soeda, the cost of developing this sequencing plant, named HUGA 1 by its designers, was roughly $10 million and the intrinsic cost of the machines is close to $1 million. Note that this project was originated by Akiyoshi Wada and launched in 1981, and had stimulated some interest [9] before apparently dropping into oblivion. This achievement highlights two salient features of research in Japan: ability to emphasize the long term, and success in getting many firms (who are also competitors in the market) to work together on a common project. In the next phase, operation of the assembly line will be handed over to private companies who will carry out sequencing under contract; the Riken team will then start working on a second generation set-up. This is how the potential conflict between sequence "production" and method improvement, that considerably disturbed megasequencing laboratories, is tackled here.

... and the Généthon Mark II room

The second example lies in a Paris suburb, in the "Ville nouvelle" (new town) of Evry. The Généthon is an *ad hoc* organization set up by AFM (French Muscular Dystrophy Association) and CEPH (Centre for the Study of Human Polymorphism). Its aim is to stimulate research on human genetics by providing the scientific community with a pool of major resources (equipment and operating staff) so that tasks can be accelerated at certain critical times. This is one way of centralizing the implementation of robots and of facilitating their management.

The most imposing part of Généthon is the Mark II room housing a series of Southern blot robots and their related equipment (Figure 3-5, p. 29). In the application of genetic mapping to localize the gene of a hereditary disease, the preparation of blots used to analyze RFLPs in families [6] is a bottle-neck. AFM therefore decided to have 20 Southern blot machines constructed; these were exact copies of the second prototype developed by the Bertin Company in collaboration with CEPH as part of the European "Labimap" project, costing about 1 million FF (200,000 US dollars). Each machine can load and then run 16 gels simultaneously, then perform the transfer onto nylon filters by electrophoresis (without any other manipulation); each has its own power supply, cooling system and controlling microcomputer. The complete instrument is large in size but development delays have been avoided. Moreover, the independence of each system facilitates maintenance and servicing. The machines are operational and most run on every working day: half for the specific needs of CEPH and Généthon (including Daniel Cohen's whole-genome contig building project [1]) and the other half for external teams. One of these, Arnold Munnich's group at the Necker Hospital has thus obtained 920 Southern blots in a matter of days. Bottle-necks do still persist, notably in the tricky manual preparation of the gels and perhaps also in the subsequent hybridization. All this is understandable when making such a two-fold jump in magnitude: the installation can produce 320 blots at a time...

A number of specific scientific projects, already described in preceding chapters, are carried out at Généthon: YAC contig building, which recently resulted in the complete physical map of human chromosome 21 [2] simultaneously with the Y chromosome map [3], whole-genome physical mapping [1], massive cDNA sequencing, and construction of a second-generation genetic map exclusively based on microsatellites makers [10]. Very important points have been scored, and Généthon has greatly improved the rank of France in Genome research. Its creation was only possible because of the existence of AFM, a powerful association of patients with large financial resources, capable of taking technological risks (as well as financial ones since the total yearly budget for Généthon is of the order of 70 million FF, close to 15 million US dollars), that would undoubtedly scare off study sections of INSERM or CNRS. This installation has also largely benefited from the initial work carried out for CEPH and from the LABIMAP programme that had produced operational, if not fully developed, machines.

Some (initial) lessons from these experiences

The systems we have discussed above are clearly special in terms of the amount of capital investment, the centralized organization, the tenacity exhibited at Tsukuba and the capability of taking large financial risks, in the case of Généthon. Furthermore, both integrate robotics based on existing methods with the innovations focusing on their automation and not on their principle. Given the development time, these systems cannot incorporate the latest technical novelties. At Tsukuba, sequencing is based on the very conventional "m13 shotgun" strategy, i.e. random DNA fragmentation, subcloning in the m13 phage and collection of sequences until they overlap enough for assembly, without any current use of PCR amplification. Similarly the Généthon Mark II machines automate the Southern blot, a commonplace, if not obsolescent, operation. There is a tendency by some, especially in the USA, to look down on this kind of automation and to advocate the development of new methods specifically adapted to robotics, rather than the mechanization of existing techniques. These reactions probably contain an element of pique...

These experiences cannot be generalized as they took place under very specific conditions. What can be hoped is that they will promote wider use of the items which can most easily be integrated in a conventional laboratory, i.e. the commercial Southern blot machine derived from the Mark II, marketed in late 1992, or some elements of the HUGA assembly line beyond the already marketed Seiko robot and Hitachi sequencer. In all cases, this will require long-term effort, a search for first-class partners and a close watch on the quality of the key men placed at the interfaces. Hopefully, it will then be possible for researchers to devote more time to mental reflection and to *researching*, instead of toiling away at their Eppendorf tubes. The professionalism which is both induced by and necessary to automation will be welcome in our laboratories, in which work sometimes seems exceedingly amateurish.

REFERENCES

1. Bellanne-Chantelot C, Lacroix B, Ougen P, Billault A, Beaufils S et al: Mapping the whole human genome by fingerprinting yeast artificial chromosomes. *Cell* 1992 **70**: 1059-1067

2. Chumakov I, Rigault P, Guillou S, Ougen P, Billault A et al: A continuum of overlapping clones spanning the entire human chromosome 21q. *Nature* 1992 **359:** 380-387
3. Endo I, Soeda E, Murakami Y, Nishi K: Human Genome analysis system. *Nature* 1991 **352:** 89-90
4. Foote S, Vollrath D, Hilton A, Page DC: The human Y chromosome. Overlapping DNA clones spanning the euchromatic region. *Science* 1992 **258:** 50-66
5. Jordan B: Heurs et malheurs du séquençage à grande échelle. *Médecine/Sciences* 1991 **7:** 612-613
6. Jordan B: L'art et la manière d'étudier les génomes. *Biofutur* 1990 **94:** 10-20
7. Maxam AM, Gilbert W: A new method for sequencing DNA. *Proc Natl Acad Sci* USA 1977 **74:** 560-564
8. Sanger F, Nicklen S, Coulson AR: DNA sequencing with chain-terminating inhibitors. *Proc Natl Acad Sci* USA 1977 **74:** 5463-5467
9. Wada A: Automatic DNA sequencing. *Nature* 1984 **305:** 193
10. Weissenbach F, Cyapay G, Dib C, Vignal A, Morissette F et al: A second generation linkage map of the human genome based on highly informative microsatellite loci. *Nature* 1992 **359:** 794-801

10
Genome research in the top two: from Livermore to Tsukuba

A very exclusive club

It will surprise no one to hear that serious genome research is carried out almost exclusively in the developed nations. Hence little will be said here about the work carried out in Central Africa or even in South America – though laboratories in these regions can play a role in selected fields, particularly for clinical aspects. Likewise, the contributions of the ex-USSR and the former Eastern bloc countries are still very modest despite recent efforts [6]. In fact the main protagonists are the USA, the EEC and, far behind, Japan; within Europe, Great Britain is the unchallenged leader with France a distant second. This ranking is based on a series of indicators which turn out to be quite consistent (Figure 2-4, p. 14). The first, and perhaps most objective, is the number of publications in the field, e.g. as listed by the Academia Europea in a 1991 study [1]; a second indicator is the percentage of scientists coopted by HUGO for each country. Another possible criterion is the number and quality of oral presentations at the "Genome mapping and sequencing meeting", held annually since 1988 at Cold Spring Harbor, and the number of speakers at this symposium – although a certain bias in favour of Anglo-Saxons, particularly those residing in North America, must be taken into account. Irrespective of the criterion, the order remains unchanged, with the USA a long way ahead, Great Britain a well placed second, and France and Japan neck and neck but quite far behind. It is interesting to note that the order does not correspond to the economic power of these nations, showing the importance of cultural factors.

In the USA: a robust genome Programme

The genome Programme in the USA is alive and well, firmly rooted in the country's scientific fabric. Abandoning or even significantly slowing down this project seems unthinkable, even in the turmoil following the resignation of its undisputed leader, Jim Watson. This was certainly not the case at the end of the eighties. The leardership of the project is entrusted to the Department of Energy (DOE) and to the National Institutes of Health (NIH); the National Science Foundation (NSF) is not very much involved in biological research. The laboratories working on the programme range from the academic university team whose research foci are very precise biological problems (generally the isolation of disease genes), to the "mapping factory" whose sole aim is to establish physical and genetic maps, using machines and operating according to an industrial-type timetable.

Let us first discuss each of the agencies that fund and coordinate this research. DOE enjoys an annual $60 million genome research budget, which supports three main centres: Lawrence Livermore (primarily on chromosome 19 under Anthony Carrano); Los Alamos (chromosome 16 under Bob Moysis); Lawrence Berkeley (chromosome 21, until late 1990 under Charles Cantor but now led by Jasper Rine, professor of genetics at the University of Berkeley). Smaller teams working in the DOE laboratories of Argonne, Brookhaven and Oak Ridge also benefit from this funding. A quarter of the budget goes to external groups, such as George Church at Harvard, or even to foreign teams like Grant Sutherland in Adelaide (Australia) who is collaborating on the genetic and cytogenetic map of chromosome 16 with the Los Alamos Centre. Out of its annual budget of just over $100 million, NIH finances nine centres: Rick Myers (UCSF, San Francisco for chromosome 4); David Schlessinger (Washington University, Saint-Louis in a centre using YACs to physically map chromosomes X and 7); Glen Evans (Salk, San Diego for 11); Ray White and Ray Gesteland (University of Utah, Salt Lake City for genetic maps of 16, 17 and 5); Tom Caskey (Houston, for regions of X and 17); Francis Collins (Michigan University, Ann Arbor for research on disease genes); and Beverly Emanuel (University of Pennsylvania) to map chromosome 22. Each of these laboratories is staffed by 30-40 persons and receives a total budget of roughly $10 million (including salaries) over a four-to-five year time interval. In addition, a large centre in Cambridge, Massachusetts, funded at 24 million US dollars for five years and headed by Eric Lander, concentrates on large-scale, whole-genome STS content physical mapping, using the CEPH "MegaYacs". Another centre in Iowa will focus on the generation of large numbers of microsatellite markers and on genetic mapping.

The sociology of "Genome Centers"

Although some of these centres opened fairly recently with grants that were awarded in late 1990 or early 1991, dissimilarities were already evident during my visits. For instance at Saint-Louis, a highly structured organization has been set up so that physical maps of chromosomes 7 and X can be methodically established from one end to the other by assembling YACs into contigs. David Schlessinger and his colleagues attach great importance to doing this in a complete and systematic fashion. Their strategy consists of defining on each of these chromosomes one or two thousand STS, procured by sequencing previously known probes or clones from chromosome-specific libraries. The YACs are then obtained from a general library by PCR-screening using these STSs, and are assembled into contigs by analysis of their "STS content" [10] – i.e. two YACs that contain the same STS must overlap since each of these markers is unique. This laboratory gives the strong impression of wanting to get to the bottom of things and of not intending to be distracted by the study of a specific region, however tempting it may appear. In contrast, Tom Caskey in Houston has explicity been awarded his grant to bolster service activities (e.g. YAC screening, cytogenetics, sequencing, data processing, etc.) for a series of teams, each of which is attempting skillfully and resolutely to isolate the gene responsible for one or two diseases.

Clear-cut disparities also exist between DOE and NIH. DOE is more "professional" and more technologically oriented; the staff in its Genome centres is more permanent and the competitive climate is somewhat less intense than in the academic environment. This appears to provide an adequate setting for laboratories undertaking the complete study of a chromosome, such as Lawrence Livermore and Los Alamos. Few post-docs and even fewer students are seen in these teams which deploy massive technology and run long-term programmes. Good results are obtained, particularly as YACs arrived just in time to supplement a physical mapping strategy, which had been devised back in the era of cosmids and was beginning to reach its limits. More modern, completely YAC-based physical mapping projects are however providing very strong competition, as discussed in Chapter 5...

Though very disparate, the NIH Genome Centres do have the common characteristic of being close to the university community. This makes for dynamic laboratories, full of life, with eager students and post-docs, and with a healthy – though sometimes excessive – turnover of personnel; but it is more difficult in this setting to remain systematic. The urge to explore focused and promising offshoots of the general project is strong, and may lead to abandoning the necessary methodical enterprise. This is a fundamental dilemma that is difficult to resolve: the justification for the genome programme is precisely to foster such a global approach, whose necessity is however not sufficiently recognized in normal university circles – nor in grant study sections.

Lessons learnt: maps are feasible

Genetic mapping has progressed significantly to the point that it is now realistic to aim for a map featuring a polymorphic marker every two or three centiMorgans on average

[12] – an objective that is almost a reality for some regions [17]. Nonetheless, the most significant headway has been recorded in the establishment of complete physical maps, based on cloned material (cosmids and especially YACs) and covering the entire chromosome. These maps are now well advanced for chromosomes 7, 11, 16, X, 19; complete maps have been obtained for the Y chromosome [9] and for chromosome 21, which was the focus of work at Lawrence Berkeley but has been successfully tackled by a French group [8]. Dividing the work up in terms of chromosomes is not artificial in this case because of the advantages of scale gained from studying an entire chromosome in a given specialized centre, and the uniform quality of the map eventually obtained. The end-result of these studies is a physical map, together with a series of clones useful in any subsequent localized investigation. These endeavours are costly: the complete physical map of an average-size chromosome mobilizes 30 or so people in one laboratory for several years and costs $10 million using this technology; the global YAC fingerprinting strategies developed in several groups [7] may bring these figures down significantly.

In all these programmes, YACs have taken over a large share of the market from the cosmids – it might even be said that they have stolen the show. A dozen copies of the general library established at Saint-Louis have been produced and installed in each genome centre. US teams are concentrating on improving the methods of screening (all-out PCR screening or hybridization on pools of clones separated on pulse field gels) and of analysis (end-cloning by "vector PCR", homologous recombination or circularization, then production of internal probes by Alu-PCR or Line-PCR [13]), rather than on constructing new libraries. In this field Europe seems slightly ahead as new and better libraries have been constructed by Rakesh Anand [5], CEPH [3] and Tony Monaco [16]. Nevertheless other teams are still working to develop new vectors and, in particular, are attempting to propagate large segments of exogenous DNA in bacteria, which are easier to manipulate than yeast [14]; these efforts have been outlined in Chapter 4.

The USA have taken up cDNAs

This topic has already been developed in our discussions on DNA sequencing (Chapter 7). Systematic sequencing of cDNAs – in principle the strong feature of the British, French and Japanese genome Programmes – has been, in fact, most efficiently taken up by the US group of Craig Venter at Bethesda in Maryland [2]. Clones from a cDNA library were randomly sequenced with the aim of obtaining a partial sequence for each of them (one machine run providing several hundreds of nucleotides for each sample). This work generates partial sequences that only reveal part of the message contained in the gene; however, this information is still valuable and the approach is extremely efficient: an average-sized laboratory can expect to handle several thousand clones every year and hence to open up a large breach in the mountain of several tens of thousands of human genes still unknown today. The attempt to register patents for these partial sequences has sparked a lively debate whose outcome is still in the balance.

This same type of approach is being applied by many teams throughout the world ranging from the Resource Centre of the British Genome Programme, the Genexpress Programme led by Charles Auffray at Généthon, to Kenichi Matsubara's laboratory in Japan. The Americans have nevertheless acted with their customary efficiency: starting as a late runner (the DOE and NIH Programs did not include cDNA until 1990), they are now the acknowledged – but not unchallenged – leader in this field.

Data processing is taken seriously

Genome informatics receives much attention in the USA, where a typical Genome Center boasts a team of four to eight data processing experts – many more than in the largest of the French INSERM and CNRS units. The Genome Data Base (GDB) set up by Peter Pearson in Baltimore has a staff of thirty or so computer and management specialists and an annual budget of several million US dollars. Moreover, the workstation is on the way to replacing the top Macintosh model or the fast PC in laboratories and on the researcher's desk. The reasons both for this effort and for the obstacles it encounters have been already discussed, and I will just reiterate the main points: the fact that improvements to techniques and their initial automation increase the volume of the results, clones and data; that techniques are increasingly integrating computers for direct data input and subsequent data processing; and, finally, the paramount importance of archiving these results and disseminating them to the scientific community, after a well defined validation process and in accordance with a precise procedure [4]. There is no lack of tasks to be carried out and the resources allocated in the USA are equal to this task, which does not mean that everything is perfect: the well-known cultural gap separating biologists and computer specialists is still only too real, with consequent difficulties in communication. Nonetheless overall progress is being made, computerized lab notebooks are starting to be used and private, semi-private and public databases are being set up. Easy access to these libraries – a prerequisite for efficient work in human genetics – is in general possible in the USA, thanks to the availability of high performance data processing equipment and especially to the high speed data links interconnecting universities, databases and computing centres.

The overall track record is positive !

This overview highlights the robust health of the US Genome Program: it is well established, mobilizes numerous laboratories and is amply funded; it has already produced some noteworthy achievements and demonstrated that it can integrate technological evolutions (YACs, PCR, microsatellites) and can regularly reorient its objectives: a case in point being the recent emphasis placed on cDNA or the integration of the CEPH "MegaYACs" in the mapping process. This programme now enjoys widespread acceptance throughout the scientific community even though some criticism is still voiced and its new leader, Francis Collins, is widely respected. Thus the lead taken by the USA - responsible, it will be recalled, for half the world "production" in this research sector - is not close to evaporating... even though the 1992 spate of results from Généthon represents a challenge.

Genome in Japan: myth and reality

Japan, the land of fantasy and of unlimited misunderstandings... It must be confessed that for the West this civilisation is particularly impenetrable: the principles governing its organization and the practices of personal relations are obscure to us and, as a consequence, misinterpretation is always close at hand. Yet Japan is a fascinating country and a number of features of this culture gradually come into focus provided we act with humility and do not try to impose our own way of thinking. In this respect, my own survey enjoyed several advantages: I was not completely ignorant of Japanese culture (this was my fifth visit), and I had already met most of my contacts before - a "must" in this society where individual relations and the long term view are of prime importance. In addition, the advice that Kenichi Matsubara (Figure 10-1) - a key figure for genome research in Japan - was kind enough to offer helped me considerably in targeting my visits; and finally, I had visited most of the laboratories in late 1988 so I could evaluate their evolution over almost three years - a valuable landmark in relatively unknown territory.

Illusions and delusions

The history of the Japanese genome Programme is a good illustration of how misunderstandings arise with respect to this country. In the late eighties, many were firmly convinced that Japan was a serious contender in research on the human genome: the resourcefulness and determination of its scientists and its undisputed know-how in robotics already made it, as was widely believed, a serious competitor for US human genome research. This widespread impression was nevertheless wrong, based as it was on a string of misinterpretations and ambiguities.

The first misunderstanding - still current among the general public - is that the primary and immediate aim of the genome programme is massive sequencing of human DNA. Sequencing three thousand million nucleotides using existing techniques would indeed be an ideal task for the Japanese as we tend to imagine them. In our often distorted mental picture of the country, they are seen as particularly adept at repetitive and unimaginative tasks. However it is clear by now that the better part of genome programmes is devoted to the construction of physical and genetic maps. Given the constantly evolving methods, this activity requires brain power and cannot therefore be entrusted to battalions of technicians... or robots !

A second misunderstanding has also crept in, this time concerning the Human Frontiers Programme. At the outset this large-scale international programme, which awards grants for joint projects comprising teams from several countries, was financed almost exclusively by Japan. According to the general belief and even to some written statements, the aim of the programme was to study the genome: by this means, Japan was seen as striving to take over leadership of the project. Ambiguity was reinforced because the objectives were presented, as sometimes in that country, with a great deal of fanfare but an undeniable lack of clarity. Today, however, everything is sharply in focus: the Human Frontiers Programme (whose office is at Strasbourg, France, and to which other countries beside Japan are beginning to contribute) supports studies on the nervous system and on brain functions... but no actual genome research.

Last of all, the public image of Japan in this field was influenced by the efforts of Akiyoshi Wada. This famous and very influential scientist expended a great deal of energy in the early eighties to set up resources for very large-scale sequencing [19]. This programme did not in fact succeed at the time because the technology was not mature enough – as we have seen with the relative failure of megasequencing projects initiated later [15]. However the programme did produce several papers whose optimism was rather ill-advised: one, published by *Nature* in 1987 announced a sequencing cost of 17 US cents per base [20]. Five years later agencies such as DOE were still trying to assess the cost, and, as we have seen, a figure of $1 remains a rather optimistic evaluation... These publications naturally gave further credibility to the impression that Japan was forging ahead on genome research.

The reality in Japan

In fact, when I visited a number of Japanese molecular biology laboratories in late 1988, achievements in genome research seemed relatively modest. This view was corroborated by the small number of Japanese papers – often of only average quality – presented in specialized international symposiums, such as the Cold Spring Harbor Genome Mapping and Sequencing Meeting. Publication surveys conducted in 1990 and 1991 confirm that the Japanese "produced" slightly less than France in the field of human gene mapping [1].

Why should this be so, in this economically powerful and technically sophisticated nation ? There are, I think, three basic reasons. The first one is the relative weakness of medical genetics in Japan, stemming from cultural factors. Intense shame is attached to hereditary defects, much more so than in the West, and, as a general rule, persons so affected are cared for within the family and not sent to institutions. Blood sampling (the basis of all family studies) is also subject to strong inhibitions. Accessing the patients is thus time-consuming and their study difficult; this has led to underdevelopment of medical genetics, which elsewhere formed the spring board for progress in human molecular genetics and then for studies on the genome.

Another reason for this lag is the relative weakness of fundamental research in this country, which first set about developing applied – and sometimes highly applied – research programmes with great success, but only allocated appropriate budgets to fundamental laboratories many years later. Notwithstanding certain opinions to the contrary, genome studies do belong to this sector (as shown by innovation such as pulsed gels, YACs or non-radioactive *in situ* techniques) and cannot be performed by mere application of already proven methods. Furthermore, it is not possible to increase overnight the number of laboratories, and even less so of expert scientists, needed to accelerate the pace of this work.

A third and more circumstantial explanation is the number of government agencies responsible for research. Coordination of their projects is thus arduous and, in comparison, the occasional quarrels between DOE and NIH, or in my country between INSERM and CNRS, are more akin to lover's tiffs*. These interministerial feuds account for the existence today of not one but four or five genome programmes (even though their coordination is improving, as we will see later), making it especially tricky to find one's way through such complex scenery.

Things get underway

But in Japan as elsewhere, no situation remains frozen for ever. Indeed genomic research in the land of the rising sun has progressed very markedly, as I found out in my discussions with the principal leaders in the field and in my visits to their laboratories. A large part of the credit for this belongs to the organizational work carried out by Kenichi Matsubara (Figure 10-1), a reputed scientist and high-class organizer. He directs the programme of the Japanese Ministry of Education (Monbusho), but also plays a part in the coordination with the other projects, those of the Science and Technology Agency (STA), of the Ministry of Health and of the

---

* Although I should moderate my comments on the situation in Japan given the complex and antagonistic development of the French Genome Programme (Chapter 11)...

Figure 10-1 Two generations of Japanese scientists. In the upper photograph, Kenichi Matsubara, head of the Genome Programme of the Monbusho (Ministry of Education), seen in his laboratory at Osaka in June 1991; and in the lower photograph, Fumihiko Matsuda, one of the talented up-and-coming "youngsters" in Japanese genome research. (Picture taken during the first European HUGO Conference at Frankfurt in December 1990.)

Ministry of Agriculture and Fisheries. The total annual funding of these programmes amounts to roughly 2.5 million Yen, approximately $20 million – that is, in absolute terms, almost as much as is allocated in France, though proportionally less compared to the country's economic power. Notwithstanding, this investment is still impressive *.

The initial goal of the Monbusho Programme (around $7.5 million in 1993) is to widen, systematize and coordinate the studies now in progress: libraries of complete cDNA in expression vectors (H. Okayama in Kyoto), isolating probes and genetic mapping of chromosomes 3 and 11 (Y. Nakamura, Tokyo), implementation of computer systems and access to data banks (M. Kanehisa, Kyoto and then Tokyo), and even sequencing of *S. pombe* (M. Yanagida, Kyoto) and physical mapping of *S. cerevisiae* (K. Isono, Kobe) – to mention but a few of these studies. The important point is that these are not large-scale, futuristic and somewhat abstract programmes but soundly based research projects, already pursued at a good international level by competent teams. The aim of K. Matsubara is to give these projects the ability to grow while coordinating them within the framework of a general "genomic" approach. There is no focus as yet on a particular chromosome or region. A highly pragmatic approach, in other words, that is beginning to produce results in several of the laboratories quoted above.

The other major programme, larger in fact than Monbusho's, is managed by Yoshi Ikawa for the STA [11]. The annual budget of more than $10 million supports research in the agency's own two laboratories (the Riken Life Science Center at Tsukuba and the Institute of Radiation Biology at Chiba) as well as funding university teams through grants – which provide very welcome financial assistance. As its name indicates, STA emphasizes technological aspects and it was under its patronage that the first proper Japanese YAC library was set up; in fact T. Imai (now in the Tokyo laboratory of Y. Nakamura) had started constructing this library in the USA at Saint-Louis and then continued in the Riken Center at Tsukuba. The Riken Center is also working on the automation of DNA sequencing (the continuation of Wada's programme), which we mentioned previously when discussing instrumentation. The laboratory at Chiba – not to be confused with the future Sequencing Institute, which is funded by the local authorities and industrialists – focuses mainly on *in situ* hybridization. In this way, it has positioned several hundred cosmids on chromosomes 3, 11 and 21 [18], those on which the STA is concentrating. As we mentioned above, STA grants are vital to university laboratories, which often receive more money from this source than from Monbusho. In the past some of the STA programmes lacked credibility, and the results obtained were not

---

* The widely varying accounting procedures mean that any comparison between different countries is very approximate and provides only an order of magnitude.

in keeping with the stated ambitions. Considerable headway has since been made, the training of research workers has borne fruit and high quality work is now being performed. STA is also responsible for setting up the Japanese node of GDB (Genome Data Base).

The remaining programmes are those of the Ministry of Health (of the order of $5 million for studies targeting selected genetic diseases), of the Ministry of Agriculture and Fisheries (with equivalent funding and including a Rice Genome Project), and of MITI (*), which stimulates industrialists into developing instruments or setting up sequencing institutes. These latter projects are less firmly established; though their immediate impact is still limited, they must nevertheless be taken into account.

Japan: a power to be reckoned with

Visiting these laboratories again after a lapse of three years proved extremely instructive: insofar as research on the genome is concerned, the outlook had significantly evolved. Several of the teams, such as that of Fumihiko Matsuda's (Figure 10-1) in the Kyoto laboratory of Tasaku Honjo, have become very competent in the use of YAC libraries. Activities such as chromosome mapping and cDNA analysis are now also well developed. Neither the programmes of the Monbusho and the STA nor the personality of Kenichi Matsubara are disputed and one has the impression that the country's involvement is serious and long-lasting.

The changes are particularly striking in terms of instrumentation. We do expect the Japanese to be impressive in this field, as they have already been with computers and video camera recorders. The reasons for this are structural: the existence of large, integrated industrial groups like Mitsubishi who enjoy in-house expertise on a wide variety of technologies; an in-built tendency to promote long-term investment; and a general preference for automation rather than recourse to foreign manpower. However the initial results in the genome instrumentation sector had seemed disappointing. Akiyoshi Wada's programme apparently got bogged down and Seiko Instruments marketed equipment automating an obsolete technique, DNA sequencing by the Maxam and Gilbert method.

These teething problems now belong to the past and Japan is currently manufacturing and commercializing advanced robotic equipment for both the domestic and foreign markets, such as a very compact Seiko robot for performing

---

(*) Strongly backed by Japanese industry, the MITI (Ministry of International Trade and Industry) is very active in programmes, such as Human Frontiers, and in the creation of advanced research institutes, such as the Protein Engineering Research Institute (PERI) at Osaka.

the sequencing reactions (naturally by the Sanger method); Fuji's imaging plate system that can replace autoradiography in many applications; a Kurabo robot, developed from the US Amgen machine, for DNA preparation, capable of processing 160 samples without assistance; and the astonishing "DNA sequencing factory" at the Riken Centre (Tsukuba), which Eichi Soeda showed me. It will be recalled that this plant (described in Chapter 9) is an assembly line with about ten robots, each of which handles a given operation: sampling the colonies of bacteria, preparing the DNA, sequence reactions, pouring the gels, etc., up to and including disposal of the gels at the end (Figure 9-1, p. 106). The line not only includes the robots but also the transfer systems to move samples from one machine to another. It still remains to be seen how the system will operate, how reliable it will be and whether its true output will match the impressive estimates, but, to the best of my knowledge, both the project and the system are unique in the world.

A breakthrough of Japanese instrumentation is therefore likely, particularly because the country's industrialists are applying a strategy that has proven successful in the past: making high-cost instruments previously produced in small numbers suitable for the mass consumer market. For instance, Hitachi, which currently sells a DNA sequencer comparable to the LKB/Pharmacia model (at a comparable price), is reported to be well advanced in the development of a much cheaper "personal sequencer" for widespread commercialisation. A large potential market certainly exists for such a machine as most molecular biology groups would be interested if the price was low enough. All these technical developments are first tested in Japanese laboratories – who will thus be the first to benefit from them. In the future, Japan may become one of the leaders in genome research, a new factor that must be taken into account and that ought to encourage us to undertake more cooperative ventures. Though relations are more complex to handle than with Western partners, for both geographical and cultural reasons, there are a number of opportunities here that it would be wise not to miss.

REFERENCES

1. Academia Europaea: Research on the Human Genome in Europe and its Relation to activities Genome in Europe and its Relation to activities elsewhere in the world 1991
2. Adams MD, Kelley JM, Gocayne JD, Dubnick M, Polymeropoulos MH et al: Complementary DNA sequencing: expressed sequence tags and Human Genome project. *Science* 1992 **252**: 1651-1656
3. Albertsen HM, Abderrahim H, CannHM, Dausset J, Le Paslier D, Cohen D: Construction and characterization of a yeast artificial chromosome library containing seven haploid Human Genome equivalents. *Proc Natl Acad Sci USA* 1990 **87**: 4256-4260
4. Aldous P: Human Genome databases at the crossroads. *Nature* (News) 1991 **352**: 94
5. Anand R, Riley JH, Smith JC, Markham AF: A 3.5 genome equivalent multi access YAC library: construction, characterisation, screening and storage. *Nucleic Acids Res* 1990 **18**: 1951-1956

6. Bayev AA: The Human Genome project in the USSR. *Biomed Sci* 1990 **1:** 106-107
7. Bellanne-Chantelot C, Lacroix B, Ougen P, Billault A, Beaufils S et al: Mapping the whole human genome by fingerprinting yeast artificial chromosomes. *Cell* 1992 **70:** 1059-1068
8. Chumakov I, Rigault P, Guillou S, Ougen P, Billaut A et al: A continuum of overlapping clones spanning the entire human chromosome 21q. *Nature* 1992 **359:** 380-387
9. Foote S, Vollrath D, Hilton A, Page DC: The human Y chromosome. Overlapping DNA clones spanning the euchromatic region. *Science* 1992 **258:** 50-66
10. Green ED, Mohr RM, Idol JR, Jones M, Buckingham JM et al: Systematic generation of sequence-tagged sites for physical mapping of human chromosomes: application to the mapping of human chromosome 7 using yeast artificial chromosomes. *Genomics* 1991 **11:** 548-564
11. Ikawa Y: Human Genome efforts in Japan. *FASEB J* 1991 **5:** 66-69
12. Jordan B: Les yeux plus gros que le ventre? *Médecine/Sciences* 1990 **6:** 576-579
13. Jordan B: La montée en puissance des YACs. *Médecine/Sciences* 1990 **6:** 470-472
14. Jordan B: Des vecteurs de clonage à la pelle. *Médecine/Sciences* 1991 **7:** 503-504
15. Jordan B: Les heurs et malheurs du séquençage d'ADN à grande échelle. *Médecine/Sciences* 1991 **7:** 612-613
16. Larin Z, Monaco AP, Lehrach H: Yeast artificial chromosome libraries containing large inserts from mouse and human DNA. *Proc Natl Acad Sci* USA 1991 **88:** 4123-4127
17. Petersen MB, Slaugenhaupt SA et al: A genetic linkage map of 27 markers on human chromosome 21. *Genomics* 1991 **9:** 407-419
18. Takahashi EI, Yamakawa K, Nakamura Y, Hori TA: A high-resolution cytogenetic map of human chromosome 3: Localization of 291 new cosmid markers by direct R-banding fluorescence *in situ* hybridization. *Genomics* 1992 **13:** 1047-1055
19. Wada A: Automatic DNA sequencing. *Nature* 1984 **307:** 193
20. Wada A: Automated high-speed DNA sequencing. *Nature* 1987 **325:** 771-772

11

The Old World is still in the race

Great Britain: a bread and butter genome Programme

Great Britain: fertile ground

Great Britain has a strong tradition of research in human genetics. The quality of its scientists and clinicians is recognized, its state-run health system has considerably facilitated epidemiological studies. Furthermore the country is the birthplace of molecular biology. It was responsible for the development of protein, RNA and DNA sequencing [27] and for the discovery of the DNA structure [29]. In addition, modern methods for the study of genes and of their regulation originated as much in the United Kingdom as in the USA. Notwithstanding the intense and privileged relationship existing between the British and their transatlantic cousins – and despite an oft-deplored brain drain – the result is not a colonial-type dependence. While many researchers do leave for the USA, many also come back home and the ties forged help both parties. In spite of moderate funding, British biological research fares very well and its quality-price ratio is unquestionably excellent. It therefore appears especially worthwhile to compare its results with those of the major world powers, the USA and Japan.

The Human Gene Mapping Project: unpretentious and pragmatic

In early 1989, Great Britain officially announced its genome programme, called the Human Genome Mapping Project (HGMP). For this purpose the Medical Research Council (MRC) was allocated £11 million over three years, followed by an annual subsidy of £4.5 million to be incorporated in the MRC budget baseline. The stated priorities centred on genetic and physical mapping of "interesting genome regions", cDNA sequencing, model organisms (mouse and the nematode) and methodological improvements. The project was to be implemented by setting up a resource centre, designed to help the teams by managing collections of probes, YAC libraries, etc., and through a series of grants to existing laboratories, awarded under a "Directed Programme of Research" [2].

The programme was defined after hard and laborious discussions. The leadership was, it seems, disputed between the MRC, represented in this matter by Sydney Brenner, and the Imperial Cancer Research Fund, directed by Walter Bodmer and already quite involved in genome research. The matter was settled in favour of the MRC by the Minister of Education and Science; Tony Vickers, a scientist who had already spent several years in the administration of research, was appointed Director and set about organizing the project. The initial grants were awarded in 1989, the Resource Centre started to operate in 1990 and by spring 1993 the British Genome News Letter, first called G-String and then G-Nome News, was publishing its thirteenth issue.

The Resource Centre: a services pool for laboratories

One of the original features of the British Genome Programme is the inclusion of a Resource Centre, a central services laboratory whose aim is to facilitate the work of outside teams: the objective is somewhat similar to that of the Généthon, set up by the French Muscular Dystrophy Association (AFM), though on a smaller scale. Located in the western outskirts of London, the Centre has a staff of approximately 20 and receives annually about £1 million, including salaries: its size and funding are of the same order as a Genome Center in the USA, even though the objective is somewhat different. The Centre's five departments handle data processing, cDNA, YACs, probes and primers, and mouse genome mapping. All services are available on a no-charge basis to the 500-odd registered users, who are expected to reciprocate by feeding back information. The staff is young, with department leaders at the Ph.D. level; the Director of Biology, Ross Sibson, has spent several years working in industry, with Amersham plc, the well-known labelled molecule and reagent supplier. Tony Vickers, replaced in 1993 by Keith Gibson, oversaw the whole programme and followed the activities of the Centre closely.

The data processing team is pursuing multiple aims: training of the users, network integration of the computing resources, connection into the international networks and development of specialized software. The group is very user-oriented and it is gratifying that its "political" choices are not sectarian. Its Head, Francis Rysavy, wishes for instance to offer access to the French Genatlas database and believes that the future of a database should depend on its look-up rate and not on any *a priori* bureaucratic decision. The cDNA group has focused on partial sequencing of a very large number of these entities using the "signature" approach (300-400 bases of 99% reliable sequence on each clone), made popular in the USA by Craig Venter

[1] but actually proposed several years before by Sydney Brenner. In its first year, this small department was able to process 6,000 clones that provided 1,500 sequences, half of which were new. The YAC group offers a screening service for both the Saint-Louis and Rakesh Anand's * [3] YAC libraries; the Probe and Primer Department manages a collection of probes and provides (free) primers; last of all, the Centre is a partner in the European venture to map the mouse genome by interspecific cross-breeding and offers a probe localization service based on this method.

In conclusion, the Resource Centre is a relatively small organization which however appears solid and, above all, in direct contact with users and their needs. It is perhaps a pity that it is not (for the time being) more involved in ambitious approaches, such as those developed by Hans Lehrach (Chapter 5 and below). This is probably the price that has to be paid for its aspiration to react quickly to the requests of teams without imposing a pre-defined strategy.

A dense and dynamic research fabric

We are obviously not going to attempt to describe such a multi-faceted subject as molecular genetics research in Great Britain; we will discuss a few cases as examples to highlight its specificities.

Hans Lehrach (ICRF, London) and his "reference libraries"

The Imperial Cancer Research Fund (ICRF) is a private foundation that collects considerable sums of money: about £40 million annually – more than AFM in France – and employs close to 1,700 persons of which at least a thousand are directly involved in research. Its main laboratory at Lincoln's Inn Fields in London houses close to 50 teams, several of which are working on the genome. Hans Lehrach's group is in fact the closest to our topic. Lehrach is an Austrian scientist who has spent a great deal of time thinking about experimental strategies and assessing as quantitatively as possible their performance and reliability; he has invented a set of original methods, quite unlike those applied elsewhere.

It was his laboratory that developed the reference library concept, which we mentioned in connection with physical mapping strategies (Chapter 5). In this method, it will be recalled, DNA libraries are exploited by means of high density filters containing 10,000-20,000 clones, set out in a regular array by a

---

* This library still seems to hold the Blue Ribbon for the lowest percentage of chimeric clones (< 5%).

robot. Identical sets of filters are sent to the outside laboratories, which can then hybridize them with their selected probes, locate the positive colony (-ies) and send the coordinates back to London in order to obtain the corresponding clones (Figure 5-3, p. 59). In this process, hybridization – the trickiest step – is carried out by the outside laboratory, highly motivated to obtain the clone, and best placed to know the peculiarities of the probe used. Moreover the scheme ensures natural collection of data within the London team, through the communication of coordinates that are required to make the clones available. Additional information is then obtained by hybridizing the same filters with complex probes (clone pools, total cDNA, etc.) or multiple-hit probes (oligonucleotides) to yield as much experimental data as possible [20].

When this method is compared with the Sequence Tagged Site (STS) technique, commonly practiced in the USA, one cannot fail to be struck by its advantages. The reference library strategy replaces the STS by much more directly usable clones, while providing quite naturally for centralization and storage of results as they are obtained; moreover, the approach is useful to the other teams during its development and not solely after it has produced a complete map. This approach is extremely effective – provided all the planned experiments can really be carried out on a routine basis. It is on this aspect that Hans Lehrach's strategy will be judged: the techniques in question are tricky and, according to some experts, not "robust" enough for routine use every day. The team has recently scored quite a few points, e.g. by demonstrating the validity of the fingerprinting method for clone assembly by hybridization on a high density filter with a set of oligonucleotides [12] and by assembling YAC contigs along the whole 15 megabase genome of the fission yeast *Schizosaccharomyces pombe* [21]. It has also succeeded in extracting chromosome 21-specific YACs from a whole-genome library [26], but in terms of actual physical mapping of human chromosomes the more centralized approach implemented at Généthon has proven more successful [5, 9]. This group of about thirty persons performs a considerable amount of work and has a total annual budget slightly below £1 million – a sizeable sum, but less than a US Genome Center.

Cosmids in England: the Nematode...

The story of the Nematode *(Caenorhabditis elegans)*, to which Sydney Brenner devoted much effort in the past twenty years, is by now well known. This small worm is not normally our "cup of tea". Nevertheless this organism whose genome spans roughly 100 megabases – comparable in size to an average human chromosome – provides an excellent testing ground for genome mapping and sequencing strategies. In 1984 Alan Coulson and John Sulston started working on its complete physical map [10]; instead of being satisfied with a restriction map, they set out on the then heroic task of systematically analyzing cosmid overlaps and thereby obtaining a "contig" (a series of contiguous cosmids), that entirely spanned the genome.

The physical map is now virtually completed [11]. Cosmids were assembled using the fingerprint method invented by Sulston; the inevitable "gaps" were then filled by YACs isolated from a library constructed by Bob Waterston at Saint-Louis. In all 17,000 cosmids have been studied and their overlaps analyzed by a computerized system created by Sulston. As for the YACs, they served as probe for the cosmid library and vice versa to build bridges linking up the contigs. All this was essentially executed by hand: in other words, a "cottage industry" job compared to the resources and staff in the DOE centres, nevertheless very well organized and that reached its goals. The nematode community, it should be noted, is much more cooperative than that of human genetics and contacts were numerous, confident and frequent, with free interchange of probes, clones and information.

The laboratory is now beginning to attack the sequencing of the nematode genome with ambitious but realistic objectives: three megabases of sequence in the centre of chromosome 3 (a region rich in genes) over three years, jointly with Bob Waterston's group, with the long-term goal of sequencing the whole of the worm's 100 megabases. This second-generation megasequencing programme, described in Chapter 7, appears quite successful, unlike some of the preceding attempts [15].

... and chromosome 11

In the same technical sphere but in a different place, and this time on human material, the team of Peter Little (Imperial College of Science, Technology and Medicine in London) is methodically analyzing the short arm of chromosome 11. Cosmids from a library constructed using a somatic hybrid are analyzed by a procedure very much like Coulson's except that several steps are automated by the ubiquitous Beckman Biomek robot. The programme is run in much the same way as similar ventures in the USA (at Lawrence Livermore and Los Alamos) for the detection of overlaps and the growth of contigs, using – as is now seen to be necessary – YACs to complete the map. However, the work at Imperial College only involves a staff of three or four and less than £200,000 per year; once again the quality-price ratio of British research appears excellent.

Edinburgh prepares for the post-gene era

The MRC Human Genetics Unit in Edinburgh is one of the major MRC units, with such well known names as Nick Hastie, Howard Cooke, David Porteous, Robin Allshire and others. It may perharps be remembered that Edwin Southern used to work there (before moving on to Oxford) in a team known in the late sixties as the "Mammalian Genome Unit" – probably the first laboratory to use in its denomination the word "genome", that has really caught on since !

The Unit has a staff of about 200 and its overall focus is largely on the mouse model, gene targeting and in general functional studies; however, it remains in contact with clinical practice and is also up to date in terms of the most recent molecular technologies. Among other achievements, it has developed very high performance gene targeting vectors, a library of chromosome 11 YACs, as well as a set of X chromosomes progressively shortened by inclusion of a telomere, an extremely useful tool for mapping purposes. The Unit also houses an amazing "depository", the MRC Human Genetic Registry containing close to 5,000 registered individuals for whom over 1,000 doctors or administrative staff supply data every year. Two ladies register – by hand and on paper – long family trees, sometimes grouping together over 100 individuals with all sorts of particulars on each person. It is without any doubt an invaluable working aid, managed with a typically British sense of pragmatism; but what would the formidable French commitee on data processing and individual rights * have to say about this ?

This Unit is undoubtedly one of the best in the MRC. Though less in the limelight than an institution such as ICRF in London, or than the Laboratory of Molecular Biology at Cambridge, it nevertheless merits recognition. Not to be overlooked is that Adrian Bird – whose authoritative work on the CpG islands is well known [16] – is now back in Edinburgh after several very productive years at the Institute of Molecular Pathology in Vienna.

"Rule Britannia" ?

Is this undisputed dominance, as during the most glorious epoch of the British Empire ? At least, it can be reasonably said that Great Britain has very successfully completed the startup of her genome Programme. Despite a moderately sized budget and often austerely equipped laboratories in which the use of every microlitre of enzyme is weighed up beforehand, very high quality work is frequently produced. Intelligence (the density of brain power per square yard is high), pragmatism, originality (a much appreciated personal trait on the other side of the Channel) blend to form a highly efficient cocktail. This survey prompted me to modulate somewhat the favourable impressions left by some of the major US laboratories and to recognize that it is possible to do as well, or even better, with more limited resources.

---

* In fact, the "Commission Informatique et Libertés" operating in our country only deals in principle with computerized files; the register described here is probably kept by hand precisely to avoid complications with similar British watch-dogs.

France: a large potential and a complex situation

An honourable third place

According to quantitative criteria referred to previously, France ranks third in world research on the human genome (See Figure 2-4, p. 14); it is therefore one of the "heavyweights", with the corresponding privileges and responsibilities. Qualitative evaluation is obviously much trickier. In the highly publicized "race to the gene", several recent successes deserve a mention, such as the localization of spinal muscular atrophy by Arnold Munnich's team [23], the cloning of the fragile X region and the analysis of its variation by Jean-Louis Mandel's group [14], and the isolation of the gene responsible for Kallman's syndrome by Christine Petit and Jean Weissenbach [19] – all with help from CEPH or Généthon for genetic mapping, YAC clones and massive sequencing. It should be stressed however that in each of the above cases competition was very fierce and the French papers appeared approximately at the same time (either just before or after) as the publications of one or two other teams pursuing the same goal. Recent achievements at the Généthon [5, 9, 30] improve very significantly the competitive position of France.

France's international influence in the genome field is largely due to the existence of CEPH and to the central role the latter has been able to take in the establishment of the human genetic map and in other large-scale studies. This organization will be discussed later in detail since it is unusual in numerous respects. For the moment, let us just say that Jean Dausset foresaw in 1980 that human genetic studies would be helped enormously by assembling a "set" of families, whose DNA could be supplied to many laboratories to allow a coherent collection of data. In this matter, CEPH has and still is playing an essential role in perfecting the human genetic map while taking part in several "genomic" endeavours in association with another *ad hoc* organization, the Généthon.

One general remark is in order at this stage: French scientific research has innovated little in methodology. When the major conceptual and technological breakthroughs responsible for advances in genome studies are reviewed, it is seen that almost all of them originated outside France. Such novel concepts as genetic mapping with DNA polymorphisms [6] and the development of approaches like reverse genetics came from the other side of the Atlantic. The new tools for physical mapping, making possible maps covering whole chromosomes, were also devised outside France: pulsed field gels [28], YACs [8], non-radioactive *in situ* hybridization, "exon" trapping are essentially of American origin. The systematic approach using cosmid contigs [10], as well as irradiation hybrids, come from Great Britain; and the renewed interest in microdissection methods began mainly in Germany [21]. The only major exception is the Généthon set-up, which demonstrates creativity and daring in industrializing genome research. On the whole, the high stability found in French research organizations has not been used to make long-term investments in developing new methodology. This situation is probably due to the low esteem which technology enjoys in France and in French study sections... Regardless of the reasons, French laboratories have mostly applied – sometimes very efficiently – the techniques invented elsewhere and have innovated only on a few occasions.

A powerful associative sector

Contrary to an often held opinion, biological research in France is not an "all-state" affair. The public agencies, INSERM (a sort of NIH) and CNRS (somewhat similar to NSF but with a sizeable biology department, roughly of the same size as the whole of INSERM), with their weaknesses and strengths do, it is true, occupy a primordial place; however, private initiative plays, and has long played, a noteworthy role. This is particularly true in the case of cancer, for which two large cancer foundations mobilize considerable resources. The corresponding organization for the genome is AFM, the French Muscular Dystrophy Association. Already in existence for many years, it was revitalized in the mid-eighties by a very dynamic president, convinced of the importance of medical research.

The Association employed new fund-raising methods that were quite unusual in France at the time, notably a very successful Telethon which succeeded since 1987 in raising between 200 and 300 million FF (40 to 60 million US dollars at the 1993 rate of 5 FF to the US dollar) each year. Part of this sum is spent on direct aid to patients but the fraction devoted to research is significant: between 100 and 200 million FF. To realize what this sum represents, it must be compared with the budget of public agencies and the items in this budget analyzed. To take a concrete example, the CNRS Life Sciences department has a total annual budget of roughly 2,000 million FF; certainly an impressive amount but salaries and overheads for the staff (scientists and technicians, who are all tenured civil servants except for students and post-docs) account for 75% of this amount. The remaining 25%, i.e. 500 million FF, covers the costs of operation, equipment, real estate, etc. Since the subsidies to laboratories are not substantially changed from one year to another, the real amount available for new and free-willed actions is of the order of a few dozen million French Francs for all the biological disciplines. The figures for INSERM are similar though the staff expenses are slightly less. Thus, in comparison, the 100 million FF injected by AFM into research on genetic diseases (in addition to the operation of the Généthon discussed later) are very significant and give this foundation considerable clout...

This influence is further amplified as the objectives of an association of patients and the attitude of its representatives differ from those of the director of a public agency. The 250 or 300 million FF raised by the Telethon express a social need and plea for something to be done quickly to fight genetic diseases and to produce tangible results. It is good for scientists to be confronted with this healthy impatience; anomalies are however possible and not always easy to avoid.

Antinomical methods of organization

The French genome research community combines two extremes. On one hand, the public laboratories, usually INSERM or CNRS, of which about twenty have emerged at the international level. These institutions enjoy stable funding, but are imprisoned in the mould of the public service, with its modest support and slow decisions. At least a year and a half, for example, is required between the submission of a request to create an INSERM unit (which can be quite a small team, with just three or four Ph.D. scientists) and its actual opening, with no guarantee on the staff and funding it will be allocated. These units have very little independence (the laboratory is not a legal entity and the Director is not authorized to sign any contracts) and virtually no possibility of effective management of their staff, who are all (except students and foreign post-docs, but including technicians) tenured civil servants nominated through a complex process run at the national level.

Under these conditions it proves difficult to set up dynamic, evolving organizations, capable of reacting quickly, of emphasizing a "hot" topic and of revising priorities when necessary. It is possible nevertheless to succeed in operating efficiently but only through much effort and a few conjuring tricks. Energy thus used is not spent on research. The ingenuity expended in attempts to hire a computer scientist at the market price by means of often illegal cumulated job functions would have been better employed in perfecting experimental tactics. The action of AFM in dispensing grants and fellowships does oil the mechanism somewhat but cannot overcome the inherent defects of the system.

At the opposite end of the scale are found organizations that are very atypical (for France): CEPH and Généthon. Initially founded with the personal capital of Jean Dausset, CEPH (Centre d'Étude du Polymorphisme Humain, centre for the study of human polymorphism) remained until 1992 a private association ("Association loi de 1901" in France, a form of legalized private association for non-profit purposes commonly used by special-interest groups such as sports fishermen or butterfly collectors). Nevertheless it is largely supported by public funding, including a yearly subsidy of more than 20 million FF from the Ministry of Research; similar amounts are provided by other sources such as AFM and the EEC. Its activities are multiform and it is not always easy to distinguish between services (such as screening YAC libraries, supplying DNAs of families, etc.) and actual research work, particularly as CEPH is also involved in Généthon (see below). On the whole this organization with its staff of about 90 is well funded and operates more like a company than a public laboratory, especially in the key area of staff management: in this respect CEPH has something in common with the Genome Centers set up in the USA. On the other hand, the latter – also well funded – are in my opinion more closely controlled by their funding agencies than CEPH.

As for Généthon, it is a by-product of the "healthy impatience" I mentioned before. At the start Bernard Barataud, President of AFM, was appalled to find out that highly labour-intensive steps in the process leading to a disease gene were mostly carried out by hand. The same impression was shared by Daniel Cohen of CEPH, and led to the idea of a specialized centre, equipped with major facilities to support in-house research and provide help to external research teams. This is the concept of Généthon, that began operation in late 1990 at Evry on the southern outskirts of Paris. It represents a large-scale investment, with a yearly budget of more than 70 million FF (almost totally provided by AFM) and a total staff of approximately 130. We have already described the "Mark II hall" which contains 20 of the Southern blot machines developed in the LABIMAP programme * (Figure 3-5, p. 29; Chapter 9): by mid 1992 more than twenty external teams had made use of its services. The Généthon is also the home of several research projects, including massive cDNA sequencing with Charles Auffray, Daniel Cohen's YAC contig building both on chromosome 21 [9] and on the whole human genome [5] and Jean Weissenbach's programme aiming at the isolation of very large numbers of "microsatellites" (more than a thousand by mid-1992) to construct a second-generation genetic map [30]. Each of these projects typically involves a team of three or four scientists and engineers, a dozen technicians, and an annual budget of something like 10 to 15 million FF – means that are utterly out of reach for "public" laboratories. Doubts have been expressed [4] on the first two projects: the reliability of the CEPH MegaYACs (as faithful representations of genomic DNA) and of the maps is questioned, while it is claimed that many of the cDNAs sequenced in the Généthon project do not correspond to human sequences. Only time, and further experiment, will tell whether or not these problems are serious. The criticism levelled at MegaYACs appears excessive given that the shortcomings reported were known and had been indicated in advance.

The world of the French genome is thus very diverse. It is even a little schizophrenic and antagonistic, divided as it is between laboratories operating in a very restrictive system (much more so than a Genome Center, to use this comparison again) and others relatively unfettered and having much better resources – although less long-term security. This forms a fertile breeding ground for inevitable rivalries between teams whose funding and working conditions differ so widely. In addition this community is graced with the presence of several strong personalities, not endowed with the gift of moderation and very likely to take any criticism as a personal affront... Understandably, therefore, all is not quiet on the genome front, and the prevailing mood is not one of serenity.

---

* A European programme within the framework of "Eureka", associating CEPH with the companies Bertin and Amersham in the development of a series of controllers for molecular biology.

Is a French genome Programme necessary?

France ranks decently in the international competition and the better teams have little difficulty in obtaining grants (especially thanks to AFM). Why not just leave things as they are? In fact, a specific programme is really indispensable; AFM, which in 1990-1992 gave this research a much-needed shot in the arm, has announced its intention of switching its support towards therapeutic approaches, in particular gene therapy. In any event this assistance is far from certain: who can guarantee that the public will remain interested and that the yearly Telethon will continue to be a success? Might not another cause succeed in winning over public opinion?

Organizing a replacement solution is thus a necessity and public laboratories must be given the flexibility (in financing and staff) required to meet the challenges of genome research within the framework of a multi-annual plan. Likewise the future of CEPH (and of Généthon) must be assured by transferring them, to some extent, into a more "normal" system but without eliminating their flexible features in the process. Thus a clearly announced French genome Programme with transparent structures, means for action and funding guaranteed over several years could usefully reshape this community with its inherent contradictions and, at the same time, provide the stimulus for the new actions needed (data processing, libraries, etc.). It would also mean that France – at long last – would have a representative at the international level. Important negotiations are underway on the databases, particularly the one at Baltimore (Genome Data Base), on the patentability of sequences [18] and on other topics. The participation of France in these negotiations is wanted and expected but as long as the situation remains as complex as indicated above, no one person can speak in the country's name. Meanwhile the chance to influence the course of events is being squandered.

"GIP Genome" and "GREG": a drawn-out pregnancy

In the spring of 1990 – a long time ago already – the Minister for Research, Hubert Curien, gave Philippe Lazar, the Director of INSERM, the responsibility of drafting a proposal for a French Genome Programme. At the time, the USA had already committed considerable funds to this sector (over $100 million annually), Japan had announced its ambitions and Great Britain had the year before got its Human Gene Mapping Programme off to a good start. It was thus now or never. Philippe Lazar then tasked Philippe Kourilsky, of the Pasteur Institute in Paris, to prepare a proposal.

In his proposed plan which he drew up after consulting widely with scientists involved in the field, Philippe Kourilsky recommended the implementation of a programme with fresh funding of roughly 100 million FF per year (20 million US dollars), that would place priority on the study of genes (and hence cDNA) but that would also provide support for data processing, model organisms and technological developments. For the organization, he advocated setting up a "Groupement d'Intérêt Public" (GIP), a framework that makes it possible under French law to associate private foundations or industrialists and public agencies. Philippe Kourilsky also strongly recommended rapid initiation of the GIP, appointment of a director with an industrial management background and setting up a flexible staff hiring scheme. In a press conference held in the autumn of 1990 (October 17), the Minister unveiled the French genome Programme which quite closely followed Philippe Kourilsky's proposal (except for staff management) and was promised new funding of 50 million FF in 1991 and 100 million in 1992 [16]. The programme appeared to be off to a good start.

These hopes were to be sorely disappointed. In March 1991 – already six months after the Minister's announcement – a respected INSERM scientist, Jacques Hanoune, was charged with an exploratory mission to study how the GIP could be initiated. During the ensuing months of laborious negotiations, it became apparent that the numerous potential partners of the GIP had widely varying ideas on what should be done and how the seats on the Board of Directors should be divided up; some did not even see the need for such an organization and, even within the Ministry, support for the project was not unanimous. No headway had apparently been made by August when the 11th Human Gene Mapping Workshop was held in London. Several French scientists voiced their concern, which was loudly echoed in the press. Shortly afterwards the Minister decided in a record time to allocate research grants totalling several tens of million FF. For some budgetary reason decisions had to be made almost immediately, with the foreseeable result that grants were awarded in questionable conditions (with no public invitation to apply), thereby further aggravating tensions in an already divisive research community. Next it was the turn of the happy winners (those who had been awarded grants) to be disenchanted as the funds they had been allocated had no actual existence; in fact, they were, in bureaucratese, "programme approvals" and not "payment appropriations"...

After complex negotiations this question was finally settled well into 1992. By then a new volunteer had been found in the person of Piotr Slonimsky, a renowned yeast molecular biologist who had just retired from the management of his laboratory at Gif-sur-Yvette. Slonimsky did not initially attempt to set up the GIP, by then renamed GREG (for "Groupement de Recherche et d'Études sur les Génomes", group for studies on genomes), but instead assembled a scientific committee and, having been promised 100 million FF, launched a call for proposals which attracted many applications. A hundred grants were awarded... and actually paid, although with some delay. The GIP was finally set up formally early in 1993, and should continue to help support Genome research in France – although the level of funding may be considered too modest.

Continental Europe: the genome archipelago

Research on the genome in the remainder of Europe (excluding Britain and France) is somewhat fragmented. Undoubtedly there are excellent teams but not much is being done in the way of a systematic study of the structure of the genome; usually these laboratories become involved through medical genetics.

"Peninsularities"

Italy is the only country to have initiated several years ago a clear-cut and structured programme, mainly thanks to the impetus of Renato Dulbecco, one of the ardent supporters of the Genome Programme in the USA. Dulbecco is one of those Italian scientists who emigrate to the USA, craft a brilliant career for themselves (in his case at the Salk Institute) while maintaining close ties with their mother country, and who often end up holding a professorship at home and another in the USA. In fact they establish very helpful contacts between the laboratories of both countries. There are real "colonies" of Italians in some US genome centres: at the Baylor Institute of Genetics at Houston around Andrea Ballabio; in David Schlessinger's laboratory at Saint-Louis or even in the circle of Tony Carrano at Lawrence Livermore. These colonies can be advantageous for both sides. On the one hand, the US laboratory profits from a steady supply of post-docs to help with its research projects; on the other hand, numerous Italian researchers acquire advanced training and the teams in the Peninsula benefit from a welcome technology transfer.

Given this context, it is not surprising that the focus of the Italian Genome Programme – initiated in 1987 and funded modestly at $2 million per year – is on analyzing the distal half of the long arm of the X chromosome, a region extensively investigated at Houston and Saint-Louis. Furthermore, it also contains a gene which codes for glucose 6 phosphate dehydrogenase, one of the "classics" in Italian medical genetics. When defective, it causes a disease known as favism, studied in much depth by another well-known American-Italian, Marcello Siniscalco (Figure 11-1), who has recently set up a new laboratory at Porto Conte in Sardinia. The Italian project makes extensive use of a somatic hybrid called X 3000, which isolates this part of chromosome X in a hamster cell, and also of a YAC library specific for this region, established in the Saint-Louis laboratory and now transferred to Naples. The project involves thirty or so teams, of which a little less than ten have a real "genomic" activity: physical mapping, isolation of expressed sequences, sequencing of certain clones, technological developments, etc.

The case of Italy is unique: to the best of my knowledge, it is the only national programme focusing on a precise region of the genome. This was doubtless possible because it started from a rather low level. While Italian medical genetics enjoys quite a good reputation, "heavy" molecular biology and gene mapping were

Figure 11-1 Some of the characters in European genome research. In the upper photograph, Luca Cavalli-Sforza (glasses) and Marcello Siniscalco; in the middle, Gert-Jan Van Ommen (cap) and Andrei Mirzabekov, one of the better known Russian molecular biologists and one of the specialists in DNA sequencing by hybridization; at the bottom, Andreas Klepsch, one of the administrators of the EEC Genome Programme. (Photographs taken by the author at the 2nd European HUGO Meeting in Sardinia, April 1992.)

relatively undeveloped and the genome field was largely untouched at the start of the project. Italy does nevertheless have teams whose interests cover other regions, and even the human genome overall: the specific study of the variation in human DNA sequences now in progress [7] is due to the initiative of Luca Cavalli-Sforza (Figure 11-1) and is also stimulating intense Italian-US collaboration.

Germany: definitely reserved

Further north, Switzerland has no very marked genomic activity; however the country has excellent molecular biology groups, e.g. that of Bernard Mach in Geneva whose research on the major histocompatibility complex has a very genomic flavour about it.

As for Germany, its public opinion regards science as suspect and is very negatively disposed towards recombinant DNA and especially towards anything connected with studying the human genome (and by extension to its potential manipulation). Much of this bias is rooted in bad memories of the Nazi era, during which German scientists were largely responsible for justifying the "racial purity" programmes of the Third Reich. After the war, most of these university figures stayed on and acted as the mainstay of human genetics in Germany, and it was not until Benno Müller-Hill published his book *Murderous Science* [23] in 1988 that this question was aired openly. The suspicion aroused by this kind of research is therefore quite understandable.

An attempt to set up a concerted human genome programme was rejected by the DFG ("Deutsche Forschungs Gemeinschaft", the German research organization) in 1987; nonetheless, a programme of grants for "Analysis of the human genome using molecular biology methods" was funded later with about $5 million annually. Germany also has some top level research groups that approach this field through their work on clinical genetics, as at the "Institut für Humangenetik" in Heidelberg, directed by Professor Friedrich Vogel (author of a very well known reference book on human genetics). In addition, the first computing service centre for the European Genome Programme is located at the Deutsche Krebsforschungs Zenter (DKFZ), also in Heidelberg. This "European Data Resource for Human Genome Analysis" is given the task of facilitating user access to the various genome databases and is housed in the huge computing centre of the DKFZ, whose size and technological level are impressive. Though its objectives are similar to those of the informatics group in the British Resource Centre described previously, it is less operational having been commissioned more recently; however its project of an "Integrated Genome Database" (IGD) providing access to the various existing systems through a common, user-friendly interface is very attractive. In the European Molecular Biology Laboratory (EMBL), mention should also be made of Wilhelm Ansorge's excellent instrumentation group, developing sequencing systems, instruments and robots, mainly for applications in research on genomes. Many other teams would deserve a mention, but for the time being they are not engaged in the systematic activity typical of a Genome Center.

The Netherlands: Peter Pearson's legacy

In the Netherlands, human molecular genetics has been coloured by the strong personality of Peter Pearson, one of the first in continental Europe to focus on molecular studies applied to man. The Human Genetics Department at Leiden, which he directed during a long tenure, has taken a notable lead in the use of non-radioactive probes for *in situ* hybridization, a fast and accurate technique that was adopted all too slowly in the Old World. Peter Pearson also played a large part in organizing the Human Gene Mapping Workshops and especially in implementing computer storage of mapping data. This interest ultimately led him to leave his country to create and direct the Genome Data Base at Baltimore, recognized as the "official" library for the genome programme (Chapter 8). Now under the leadership of Gert-Jan Van Ommen (Figure 11-1), his Department is continuing clinical genetics research with massive use of *in situ* hybridization and excellent integration between clinical aspects and fundamental molecular genetics.

Other centres have teams like Ben Oostra's (Erasmus University at Rotterdam), that in 1991 took part in the isolation of the FMR-1 gene involved in the Fragile-X mental retardation syndrome (Chapter 6). Here again we find good modern genetics teams, enjoying good connections with the community, collaborating intensively with the USA and using state of the art modern methods, but doing little in the way of systematic genome work.

Scandinavia: getting organized

Despite the recent affluence of Sweden, Norway and Finland, the research fabric of these countries is limited by their small populations: Sweden, for example, has only eight million inhabitants. Nevertheless, these nations have produced high quality work in human genetics, thanks to several first rate centres (e.g. the Karolinska in Stockholm and Uppsala University in Sweden, and the Genetics Department at Helsinki University in Finland). Their studies have been promoted by sociological factors: a geographically stable population, a state-run health system and families whose members stay in close contact, all of which help the collection of data. These opportunities have motivated a "Nordic Genome Initiative" aimed at consolidating collaboration between the nations of the region through specialist workshops, improved data bank access (Uppsala is scheduled to become the third secondary GDB node after London and Heidelberg), and a common organization to be set up for genetic mapping. The latter will include distribution to twelve or so laboratories of primers corresponding to several hundred microsatellite markers.

One of the strengths of these countries is the active cooperation between research and industry. This is particularly so in Uppsala (Sweden), not only an important university town but also the site of Pharmacia's head office. The Medical Genetics Department, directed by Ulf Petterson (a leader in Nordic genome research and organizer of the Swedish genome Programme that got underway late in 1992 with a yearly budget of a million US dollars), is involved in analyses of the X chromosome, mutation studies, research on the genetic determinants of autoimmune diseases but it is also engaged in very technologically oriented work. The staff of this laboratory includes Ulf Landegren, responsible for developing the Oligonucleotide Ligation Assay (OLA) technique in Lee Hood's Department (Caltech, USA). This mutation detection method has the two-fold advantage of being easy to automate and of potentially avoiding the use of PCR [25]. Commercial utilization of the latter technique has run into licensing and royalty problems, which explains why Pharmacia is interested in this new process.

Mention should also be made of Denmark, very active in clinical genetics, Belgium and farther to the south, Spain – all countries with good laboratories that nevertheless do not pursue systematic genome studies.

The European Programme comes out of limbo

This overview of the archipelago of the European genome has underscored the quality of research in the Old World but also its relative dispersion and the need for coordination. This is the primary aim of the European Genome Programme [13], whose definition was initiated in 1987, when the USA were also beginning to tackle this subject. It had a difficult and eventful birth that underscored the political, and even psychological stakes of such an enterprise. The suggestion of a French Nobel Prize winner to entitle the programme "Predictive Medicine" was adopted to highlight its expected positive repercussions on public health. However, not enough thought was given to the misgivings that this slogan could create, with the result that the project became the target of harsh criticism from the German "Greens". Benedikt Harlin, a member of the European Parliament and an activist on this side, was involved in the committee examining the project and opposed several aspects of it. After hard and drawn-out bargaining, the project was renamed "Analysis of the Human Genome", had 7% of its budget earmarked for social, legal and ethical issues and finally got underway at the end of 1990.

Of relatively limited scope, it has been allocated an overall sum of 17 million Ecus (about $25 million) over two years, i.e. one twentieth of the US Programme. Its final goal is not exactly the same either since it supports common-interest enterprises, aimed at providing reagents or data to the European research community. The programme has six main headings: perfecting the genetic map, which includes the European Gene Mapping (EUROGEM) Programme, through which Southern blots (generated by the Généthon) are to be provided by CEPH and probes by the British Resource Centre to speed up the work of the twenty or so laboratories working on the programme; assisting physical mapping by financing the distribution of cDNA and cosmid libraries and setting up screening services for YACs; supporting the databases (with the Heidelberg Centre referred to previously); furthering technological developments: providing advanced training, and, last of all, studying the social, legal and ethical issues raised by this research work, a sector receiving funds of 1 million Ecus.

The EEC Genome Programme seems off to a good start after having been strongly criticized, initially for the time taken to set it up and then for the more or less hermetic procedures governing the invitation to tender and the evaluation of bids. Those familiar with European research grants know only too well that these reproaches are unfortunately frequent for grant programmes run by Brussels. Some scientists also resent the very strong British influence in the project administration – nonetheless understandable if one takes into account that Britain alone, in 1990 or 1991, produced as much genome research as all the rest of Europe... The second phase of the EEC genome Programme begins late 1993, with a two-fold increase of its budget.

REFERENCES

1. Adams MD, Kelley JM, Gocayne JD, Dubnick M, Polymeropoulos MH et al: Complementary DNA sequencing: expressed sequence tags and Human Genome project. *Science* 1991 **252**: 1651-1656
2. Alwen J: United Kingdom genome mapping project: background, development, components, coordination and management, and international links of the project. *Genomics* 1990 **6**: 386-388
3. Anand R, Riley JH, Smith JC, Markham AF: A 3.5 genome equivalent multi access YAC library: construction, characterisation, screening and storage. *Nucleic Acids Res* 1990 **18**: 1951-1956
4. Anderson C: Genome shortcut leads to problems. *Science* 1993 **259**: 1684-1687.
5. Bellanne-Chantelot C, Lacroix B, Ougen P, Billault A, Beaufils S et al: Mapping the whole human genome by fingerprinting yeast artificial chromosomes. *Cell* 1992 **70**: 1059-1068
6. Bostein D, White RL, Skolnick M, Davis RW: Construction of a genetic linkage map in man using restriction fragment length polymorphisms. *Am J Hum Genet* 1980 **32**: 314-331
7. Bowcock A, Cavalli-Sforza L: The study of variation in the human genome. *Genomics* 1991 **11**: 491-498
8. Burke DT, Carle GF, Olson MF: Cloning of large segments of exogenous DNA into yeast artificial-chromosome vectors. *Science* 1987 **236**: 806-808
9. Chumakov I, Rigault P, Guillou S, Ougen P, Billault A et al: A continuum of overlapping clones spanning the entire human chromosome 21q. *Nature* 1992 **359**: 380-387
10. Coulson A, Sulston J, Brenner S, Karn J: Toward a physical map of the genome of the Nematode Caenorhabditis elegans. *Proc Natl Acad Sci USA* 1986 **83**: 7821-7825
11. Coulson A, Kozono Y et al: YACs and the C. elegans genome. *Bioessays* 1991 **18**: 413-417
12. Craig AG, Nizetic D, Hoheisel JD, Zehetner G, Lehrach H: Ordering of cosmid clones covering the herpes simplex virus type 1 (HSV-1) genome: a test case for fingerprinting by hybridisation. *Nucleic Acids Res* 1990 **18**: 2653-2660

13. Ferguson-Smith MA: European approach to the human gene project. *FASEB J* 1991 **5:** 61-65
14. Heitz D, Rousseau F, Devys D, Saccone S, Abderrahim H et al: Isolation of sequences that span the fragile X and identification of a fragile X-related CpG island. *Science* 1991 **251:** 1236-1239
15. Jordan B: Les heurs et malheurs du séquençage à grande échelle. *Médecine/Sciences* 1991 **7:** 612-613
16. Jordan B: Ilots HTF: le géne annoncé. *Médecine/Sciences* 1991 **7:** 153-160
17. Jordan B: Flash. *Médecine/Sciences* 1990 **6:** 908
18. Kahn A: Faut-il breveter le génome humain? *Médecine/Sciences* 1991 **7:** 960-961
19. Legouis R, Hardelin J-P, Levilliers J, Claverie JM, Compain S et al: The candidate gene for the X-linked Kallmann syndrome encodes a protein related to adhesion molecules. *Cell* 1991 **67:** 423-435
20. Lehrach H: Hybridization fingerprinting in genome mapping and sequencing. *Genome Analysis* **1:** 39-81; Cold Spring Harbor Laboratory Press, 1990
21. Ludecke HJ, Senger G, Claussen U, Horsthemke B: Cloning defined regions of the Human Genome by microdissection of banded chromosomes and enzymatic amplification. *Nature* 1989 **338:** 348-350
22. Maier E, Hoheisel JD, McCarthy L, Mott R, Grigoriev Av: Complete coverage of the schizosaccharomyces pombe genome in yeast artificial chromosomes. *Nature Genetics* 1992 **1:** 273-277
23. Melki J, Abdelhak S, Sheth P et al: Gene for chronic proximal spinal muscular atrophies maps to chromosome 5q. *Nature* 1990 **344:** 767-768.
24. Muller-Hill: Murderous science. Oxford University Press, New-York, 1988
25. Nickerson DA, Kaiser R, Lappin S, Stewart J, Hood L, Landegren U: Automated DNA diagnostics using an ELISA-based oligonucleotide ligation assay. *Proc Natl Acad Sci USA* 1990 **87:** 8923-8927
26. Ross MT, Nizetic D, Nguyen C, Knights C, Vatcheva R et al: Selection of a human chromosome 21 enriched YAC sub-library using a chromosome-specific composite probe. *Nature Genetics* 1992 **1:** 284-290
27. Sanger F: Sequences, sequences, and sequences. *Ann Rev Biochem* 1988 **57:** 1-28
28. Schwartz DC, Cantor CR: Separation of yeast chromosome-sized DNAs by pulsed field gel electrophoresis. *Cell* 1984 **37:** 67-75
29. Watson JD, Crick FHC: Genetical implications of the structure of deoxyribonucleic acid. *Nature* 1953 **171:** 737-738
30. Weissenbach F, Gyapay G, Dib C, Vignal A, Morissette F et al: A second generation linkage map of the human genome based on highly informative microsatellite loci. *Nature* 1992 **359:** 794-801

12
Coordination or competition ?

Rivalry is the rule

The genome programme is often envisioned by outsiders as a major international project – journalists sometimes refer to "the HUGO Project" –, coordinated by authorities who define the task of each nation and even of each laboratory to avoid duplication of work and ensure maximum efficiency. This reveals a very idealized view of the world of research and of researchers themselves, seen as exemplary beings dedicated to the furthering of knowledge for the benefit of humanity. A flattering and, of course, erroneous image – just as false as the opposite cliché of the "mad scientist" perfecting, in his kitchen or garage, the fiendish invention that will make him "The Master of the World". In fact the motivations of research scientists, in genetics as in other fields, do not differ markedly from those in other professions. Over and above intellectual curiosity, the fun of performing experiments and the impression of helping in a constructive quest, ambition, hunger for fame as well as interest in material rewards play an important part.

It should therefore come as no surprise to find that the research community is competitive, sometimes brutally so. The search for the gene responsible for some disease with major scientific, economic and media importance often involves a race between several laboratories performing similar work with analogous strategies, the sole aim being to reach the winning post before the other competitors [13]. A particularly striking example of this was provided in August 1992 with four simultaneous papers describing the probable involvement of the P22 myelin gene in Charcot-Marie-Tooth disease – all published in the same issue of *Nature Genetics* [8, 11, 17, 18]... An outrageous situation perhaps since public monies or donations may be wasted in such a process, but this is how research is done in the real world. Within reasonable bounds, such competition is in fact beneficial insofar as it allows results to be checked, shares out the risks and compels laboratories to make the best possible use of their human and financial resources.

The world of the genome is no exception, subject as it is to these constraints, and it is also full of competition. This takes place between countries, with for instance a lively antagonism between the USA and Japan, accused on several occasions by Jim Watson, former director of the NIH Genome Programme, of wanting to develop commercial applications of US work without taking an active (and expensive) part in the underlying research [3, 14]. Such rivalry also exists between teams in the same country: several groups in the USA have been competing to achieve physical mapping of chromosome 21. This is a particularly appealing topic given the small size of this chromosome – just 50 or so megabases – and its involvement in widespread and serious syndromes such as trisomy 21 (Down syndrome) and Alzheimer's disease [16]. In the end, however, the first complete physical map of this chromosome was obtained by a French group [2].

The competition is nevertheless relatively subdued in the field of physical mapping: there are major advantages in approaching this work at the level of a whole chromosome; this calls for substantial resources and encourages teams to specialize. This type of study is for the time being hardly developed on a large scale outside the USA, except for the work carried out at Généthon in France. On the other hand, in all research on genetic diseases, competition is the norm, as illustrated by Duchenne's muscular dystrophy, cystic fibrosis or, more recently, the fragile X mental retardation syndrome (Chapter 6). Now and then a "consortium" forms in order to tackle a specific disorder: the most long-lasting is probably the one dealing with Huntington's chorea – now cloned at last! – that Nancy Wexler [15, 20] has coordinated for several years. However, these associations, bringing together parties with often diverging interests, are usually frail, and the inbuilt competitive strains often cause them to break up or at least to stop functioning effectively.

It is, in fact, completely unrealistic to expect that any central authority could adjudicate selected topics to a given country or to a specific laboratory and not to anyone else, though some leaders would apparently have liked to see themselves in this position [4]. On the other hand, organizations can and should be concerned with disseminating information so that the results obtained can be utilized as quickly as possible and that the contenders at least know they are in competition: it is then up to them to react by adopting complementary approaches, collaborating or simply stepping up their efforts to win the race. Existing international organizations have shown interest in becoming involved in the coordination of this research and in the distribution of its results. UNESCO, probably the most committed in this respect, has convened several international meetings, particularly on ethical issues, and has set up a system of fellowships aimed at helping researchers from the less developed countries to access the methods and results of genome research [6]. However, the most important is without doubt the well known HUman Genome Organization (HUGO), created for this cause and inspiring many hopes.

HUGO: necessary... but still searching for its identity

A blurred public image

What is the HUman Genome Organization (HUGO) ? The wide range of answers to this question is astonishing. Some see it as the all-powerful international organization running the Human Genome Programme, carving up chromosome studies between nations, coordinating strategies, handling ethical issues and providing guide-lines for the clinical and commercial applications of the newly-acquired knowledge. Colleagues, believed to be confidants of this holiest of inner sanctums, are questioned to find out how to request research funds from HUGO. On the contrary for others, it is a "gadget" ("Un machin", as General de Gaulle used to call the United Nations), one more learned assembly in which much is said but little done. This second opinion prevails in the USA; in a survey of genome research throughout the world [5], the very sober European Science Foundation has also severely accused HUGO of not doing its job.

EMBO and HUGO: very different initial conditions

HUGO was founded at the first Genome Mapping and Sequencing Meeting held in late April 1988 at Cold Spring Harbor. The initiative of a small group of big names (Victor McKusick, Jim Watson, Sydney Brenner, etc.) was approved by the participants at the meeting, who were presented an already wrapped-up project. For many, this was to be a rerun for the human genome of a successful experiment – that of the European Molecular Biology Organization. At the onset of the projects for systematic analysis of the human genome, it was indeed desirable to have an institution capable of overseeing the work of the many laboratories from several countries, improving communication, helping with training and bringing pressure to bear on the governments so that they would support the project.

Yet circumstances were quite different from those prevailing in 1969 at the creation of EMBO. The latter was European, not world-wide, and it coordinated a discipline whose centre of gravity was admittedly in Great Britain but that was also flourishing in several other European states, particularly France. In world-wide terms, the situation of the genome is quite another matter. The US share (as assessed in 1990-1991) of these activities is at least half, and considerably more when only the "heavy" genome sector – i.e. whole chromosome maps and large-scale sequencing – is considered. Though highly active, Great Britain ranks far behind with about 15% of the world output. Thus, right from the start, HUGO ran the twofold risk of either being too committed to "the Americans" or, on the contrary, of being viewed as an instrument of war against them: let us not forget that Japan is also involved and that its participation in the genome programmes can, as we have seen, loose uncontrollable passions. There were other differences. In 1969 the European molecular biology community was relatively close-knit, certainly more so than human genetics in the late eighties with its wild races, its publicity stunts... and sometimes its dirty tricks. Furthermore at the outset EMBO had three valuable assets: regular and state-independent funding from the Volkswagen Foundation; an ambitious and structuring goal, the prospect of a European Laboratory (in 1974 to become the EMBL in Heidelberg); and lastly, an excellent scientist, John Tooze, who devoted himself almost full-time to EMBO and played a paramount role in its startup. As it turned out, HUGO did not enjoy such birthday presents.

HUGO: a difficult childhood

By September 1988, the "founding fathers", whose numbers had in the meantime grown to forty-two from seventeen countries, initiated their action by a meeting at Montreux in Switzerland. They elected a Chairman (Victor McKusick), Vice-Chairmen (Walter Bodmer, Jean Dausset, Kenichi Matsubara), selected a first contingent of members and set up an election procedure, very similar to that of EMBO, to bring in about forty or so fortunate new members each year. A Board with eighteen members, all well known scientists, was formed to set down the lines of actions. However, beyond these organizational measures, HUGO encountered great difficulty in rounding up funds and effectively functioning. In late 1989, Walter Bodmer was elected Chairman, with as Vice-Chairmen Charles Cantor, Kenichi Matsubara and Andrei Mirzabekov. Annual funding of half a million dollars was provided by the Wellcome Trust and the HHMI. James Wyngaarden, a former Director of NIH, was signed up as executive director.

It was therefore reasonable to assume that HUGO was on its way and that results from its activities would soon be manifest. But, as it turned out, half a million dollars is not enormous for creating an organization (especially of international scope), finding office space and recruiting staff; James Wyngaarden, the only permanent scientist, did not prove particularly efficient, doubtless too absorbed by his other functions. The lack of a strong organization to prepare the work of the Board and to implement its decisions was sorely felt. Funding also had problems in keeping up and it turned out to be very complicated administratively for HUGO to receive subsidies from governments (assuming they wanted to allocate them). All this led to the criticisms mentioned above and to a definite disenchantment with the organization.

The stakes involved

Yet many fields could benefit from the action of an organization such as HUGO. There is no question of it running Genome programmes as a sponsor; this would require astronomical sums, whose loss of control governments would be unlikely to accept and that would be very awkward to manage in a centralized manner. Instead, the functions of HUGO should consist of coordinating, informing and advising, particularly in relation to databases and ethics. The importance of this role is reinforced by the necessary evolution of the Human Gene Mapping Workshops (HGMW), held every two years from 1973 on. The HGMWs are recognized as the grand conclaves of the map-makers: the working groups meet, chromosome by chromosome, to discuss the latest results, an agreement is reached on the most plausible version of the genetic and physical maps and, at the same time, new genes are given a logical name and the "anonymous" DNA segments are duly catalogued. All the data validated by this process can then be made available to the community in printed form, as has long been the case, in a special volume of *Cytogenetics and Cell Genetics* whose size and weight grow incessantly. Starting in the eighties these results have also been handled in computerized format, these days in the Genome Data Base (GDB) at Baltimore (directed by Peter Pearson), which is the successor to the Human Gene Mapping Library at Yale and is now the "official" database for the HGMWs.

As might be expected, the democratic and friendly system of HGMWs has been prey to the powerful pressures generated by the accelerated data acquisition rates (Figure 12-1): no longer is it possible to wait two years before updating the map, therefore "intermediate" HGMWs have been held behind closed doors (HGM 9.5, 10.5, etc.), and even the form of the "official" HGMWs is changing, not without much gnashing of teeth by the scientists who are rightly very attached to it. Nonetheless, the logistic constraints inevitably entail entering the results into databases continuously and independently of the main meetings. This process can best be implemented in chromosome-specific meetings: only a limited number of persons are required for the discussions and the amount of data is not too enormous. Such a chromosome-by-chromosome approach, with meetings staggered over one or two years, removes the need for the host of workstations and the throng of computer specialists required by the general HGMW meetings.

These specific chromosome meetings have important work to do. The probes, RFLPs, STS, microsatellites must be catalogued, the maps (genetic, physical) from the different sources scrutinized, and the positions of the newly localized genetic diseases plotted. This requires the presence of those actually contributing to the map, and the definition (and observance) of rules so that teams, that may be competing, do nevertheless provide each other and the community with as much of their data as possible. This is not always easily achieved, and for example database storage of microsatellites has been very unsatisfactory. Groups

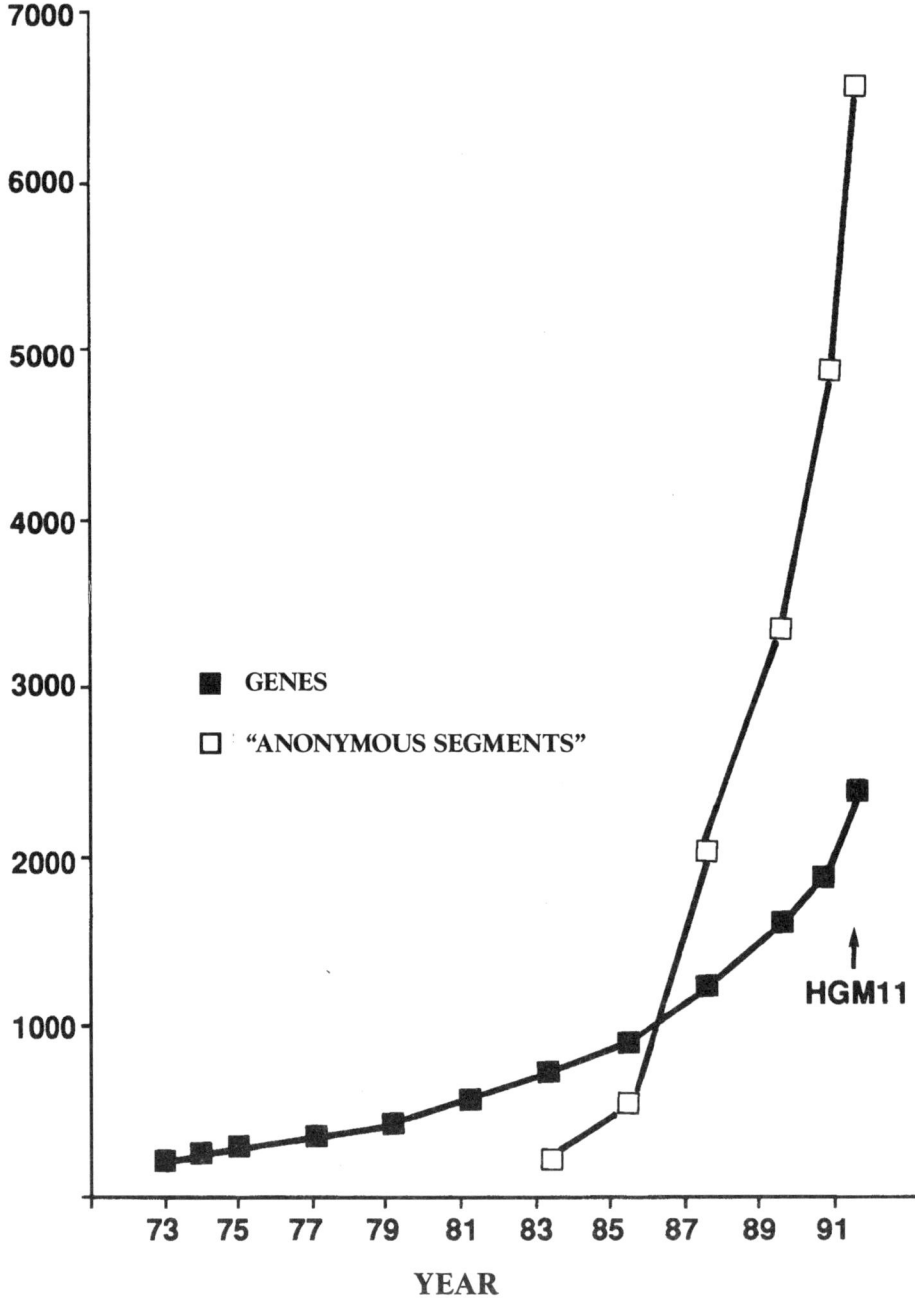

Figure 12-1 Growth in the number of genes and "anonymous" segments recorded in successive Human Gene Mapping Workshops. Note that figures are considerably underestimated due to the delay in recording the data in the public databases. The plot was provided by M. Probert, Head of the HGM 11 Data Processing Group.

constructing a detailed map of a region while searching for a disease gene did characterize many of these markers, but tended to keep them to themselves [12]. This has now improved, and large numbers of CA repeats have been made available by publication [7, 19] or in electronic media. Other considerations are also important to make these chromosome meetings successful. Their site must be equipped with suitable data processing facilities and have a good link to GDB – which does not mean they should always take place in the USA ! In short, it is highly desirable that a neutral and knowledgeable body such as HUGO should play an effective role in this organization, define some rules, make available the lessons to be learnt from other similar meetings, and ensure all those working in the discipline are well represented.

Over and beyond that, it seems clear that HUGO should be involved in managing the data coming from the various sources, which should be disseminated and be accessible. This implies agreement on a unique or compatible database(s), on the type of checks required prior to archiving the results, on the authorized confidentiality period, etc. This issue is particularly urgent for the cDNA sequences, today the goal of numerous teams and the subject of lively discussions about patentability. It is proving tricky to centralize these sequences, by definition incomplete and imprecise (they are signatures rather than sequences); however, this is mandatory if only to prevent the same entity being studied several times by different laboratories. Questions like cDNA patenting are also within HUGO's province, and the organization has in fact taken a clear stand opposing them.

The "clout" of the United States

In these different fields there is clearly a risk of domination by the USA, whose research community is dynamic and well organized while its European and Far-Eastern competitors are scattered and not particularly cohesive. Yet overall, the EEC has almost the same scientific weight, but it does not present a united front. Thus we, in Europe, find it normal for meetings to be almost always held in the USA, for the official database (GDB, funded by HHMI and then by NIH and DOE) to be in Baltimore and for the existence of a dominance that is not objectively justified, all the more as some of the strategies developed in the Old World have a very attractive quality-cost ratio. A very instructive example of this is provided by comparing landmark tactics for the physical map: the STS proposed by Maynard Olson, now more or less the gospel in the USA, the reference libraries championed by Hans Lehrach [10] in London, and the very effective YAC fingerprinting system developed by Daniel Cohen's group [1]. As we have also seen, Japan is beginning to grow in importance and the moment is ripe to include that country in a balanced international cooperative venture while it still feels the need and inclination to join it.

Naturally there is no question here of shaping HUGO into a weapon for use against the USA but, on the contrary, of using it to maintain an equilibrium that corresponds to the reality and offers the best prospect of quickly advancing the programme by giving all the various strategies a chance. We will say, like Mao Zedong in 1957, but perhaps with more sincerity "Let a hundred flowers bloom"...

Hopes for the future

Since late 1989 Walter Bodmer (Scientific Director of the Imperial Cancer Research Fund, the powerful British private foundation, mentioned while discussing that country) has been Chairman of HUGO [9]. He has taken this task very seriously and has attempted – with some success – to provide HUGO with the financial resources and, above all, the sound organization it needs. Too much was expected of HUGO – and too soon – because the obstacles in setting up an operational international organization in such a competitive sector were misjudged. Numerous hurdles must be resolved or circumvented. Nonetheless, HUGO now has an active operating office in service in London; several American scientists, notably Charles Cantor, are very active on the other side of the Atlantic (Figure 12-2) and Thomas Caskey has become president for 1993-1996. The membership

Figure 12-2 HUGO in high seas. In the center, Charles Cantor, very active on behalf of HUGO in the USA, between Elizabeth Evans, of the organization's European office in London, and the author. (Picture taken during the 2nd European HUGO meeting in Sardinia, April 1992.)

formalities for HUGO have been simplified to allow as many researchers as possible to join and to cut down the workload caused by the complex election of new members. HUGO has now set about its work of organizing special meetings for each chromosome. Its performance in this task will be the yardstick of its success.

REFERENCES

1. Bellanne-Chantelot C, Lacroix B, Ougen P, Billault A, Beaufils S et al: Mapping the whole human genome by fingerprinting yeast artificial chromosomes. *Cell* 1992 **70**: 1059-1067

2. Chumakov I, Rigault P, Guillou S, Ougen P, Billaut A et al: A continuum of overlapping clones spanning the entire human chromosome 21q. *Nature* 1992 **359**: 380-387

3. Davis B: Watson doesn't use gentle persuasion to enlist japanese and german support for genome effort. *Wall Street Journal* 1990

4. Dickson D: Watson floats a plan to carve up the Genome. *Science* 1989 **244**: 521-522

5. European Science Foundation: *Report on genome research.* Strasbourg, France, 1991

6. Grisolia S: UNESCO program for the Human Genome project. *Genomics* 1991 **9**: 404-405

7. Hudson TJ, Engelstein M, Lee MK, Ho EC, Rubenfield MJ et al: Isolation and chromosal assignment of 100 highly informative human simple sequence repeat polymorphisms. *Genomics* 1992 **13**: 622-629

8. Matsunami N, Smith B, Ballard L, Lensch MW, Robertson M et al: Peripheral myelin protein-22 gene maps in the duplication in chromosome 17p11.2 associated with Charcot-Marie-Tooth 1A. *Nature Genetics* 1992 **1**: 176-179

9. McGourty C: A new direction for HUGO. *Nature* 1989 **342**: 724

10. Nizetic D, Zehetner G, Monaco A, Gellen L, Young B, Lehrach H et al: Construction, arraying, and high-density screening of large insert libraries of human chromosomes X and 21: their potential use as reference libraries. *Proc Natl Acad Sci USA* 1991 **88**: 3233-3237

11. Patel PI, Roa BB, Welcher AA, Schoener-Scott R, Trask BJ et al: The gene for the peripheral myelin protein PMP-22 is a candidate for Charcot-Marie-Tooth disease type 1A. *Nature Genetics* 1992 **1**: 159-165

12. Pearson PL, Maidak B, Chipperfield M, Robbins R: The Human Genome initiative - do databases reflect current progress? *Science* 1990 **254**: 214-215

13. Roberts L: The race for the cystic fibrosis gene. *Science* 1988 **240**: 141-144

14. Roberts L: Watson Versus Japan. *Science* 1989 **246**: 576-578

15. Roberts L: Huntington's gene: so near, yet so far. *Science* 1990 **247**: 624-627

16. Saint George-Hyslop PH, Tanzi RE, Polinsky RJ, Haines JL, Nee L, et al: The genetic defect causing familial Alzheimer's disease maps on chromosome 21. *Science* 1987 **235**: 885-890

17. Timmerman V, Nelis E, Van Hul W, Nieuwenhuijsen BW, Chen KL et al: The peripheral myelin protein gene PMP-22 is contained within the Charcot-Marie-Tooth disease type 1A duplication. *Nature Genetics* 1992 **1**: 171-175

18. Valentijn LJ, Bolhuis PA, Zorn I, Hoogendijk JE, Van Den Bosh N et al: The peripheral myelin gene PMP-22/GAS-3 is duplicated in Charcot-Marie-Tooth disease type 1A duplication. *Nature Genetics* 1992 **1**: 166-170

19. Weissenbach F, Gyapay G, Dib C, Vignal A, Morissette F et al: A second generation linkage map of the human genome based on highly informative microsatellite loci. *Nature* 1992 **359:** 794-802
20. Wexler N S, Conneally PM, Housman D, Gusella JF: A DNA polymorphism for Huntington's disease marks the future. *Arch Neurol* 1985 **42:** 20-24

13
Genome, the trouble-maker

On several occasions, we have referred to the destabilizing effects of genome endeavours on the research community, that often operates with cottage industry practices and is not geared to the mass production of results according to a strict timetable. I will now come back to a discussion of these contradictions which highlight some of the future tendencies in biology. I will summarize the controversies which broke out when genome programmes were launched, then analyze the organizational problems as well as the ways in which scientists in charge are attempting to solve them.

A focus of misunderstandings

"Sequencing man"

The genome programme has been severely criticized, often because its aims and its organization were misconstrued. The first and foremost misunderstanding, and the source of most of the others, regards sequencing. Very respectable scientists have attacked the project on the assumption that its primary and short-term objective was to decipher the three thousand million nucleotides contained in our DNA. Given the capabilities of today's techniques, this would be, it is true, a very labour-intensive and repetitive task, it would require quasi-military organization and its scientific results would probably not be commensurate with the investment made (optimistically $1 per base, i.e. a total of $3,000 million).

All-out sequencing was indeed strongly emphasized in the original plans. The first embodiment of the genome project was conceived in 1984 by Robert Sinsheimer, at the time Chancellor of the University of Santa Cruz in California and potential beneficiary of a $36 million donation for an already funded telescope: happy country, happy times… He was trying to redirect this huge gift to an exciting biology project, that in his view centred on exhaustive sequencing of our genome. Furthermore, those presenting the project in the discussions over the ensuing years mainly stressed this aspect of the proposed endeavour [3, 8, 10, 12]. Was this a very ambitious gamble counting on fast progress being made in sequencing methods? Or a desire to announce an easily understandable objective (like "To walk on the moon") and to disregard the intermediate steps? In any event the impression had crystallized and today this misconception still persists in some circles. The readers of this book are well aware that the reality is quite different: most of the work to date has been concentrated on genetic and physical mapping, and there is actually no certainty that all of our DNA will be sequenced by AD 2005 – the approximate finishing date of the present 15-year genome programme.

Give a dog a bad name… and hang him

This misunderstanding has been the source of much tongue lashing: indeed it is easy to demonstrate that multi-megabase sequencing of human genomic DNA with today's (or even tomorrow's) technology is surely not the most formative, stimulating nor cost-effective activity in biological research. This confusion, largely responsible for fuelling the controversy, was compounded by a second error, according to which this routine, costly and ill-advised activity was depriving "true" biology of resources [7]. An examination of some figures indicates this belief is unfounded: if we consider NIH, its total budget in 1992 was approximately $9 thousand million, of which $100 million were specifically allocated to the genome programme – i.e. proportionally 1.1%. Admittedly not all of NIH money goes to what we would call biological research, allowance must be made for overheads, administrative staff and grants awarded for very applied or clinical work. Thus, in relation to the funds really available for true research, the project's impact probably amounts to several percent rather than one. This is still a modest proportion, quite out of order with the above-mentioned criticisms.

A mind-numbing routine ?

One more misunderstanding

Another misconception pertains to the internal operation of genome programmes, seen as involving essentially specialized laboratories that house armies of technicians performing highly repetitive tasks under the supervision of a few scientists. According to this quite commonly held view, the work merely consists of repeating – virtually endlessly – the standard drills and routines of a molecular biology laboratory, and the entire process is seen as particularly lacking in creativity.

This over-simplified picture is largely erroneous. Recombinant DNA technology is less than twenty years old, and is still evolving at a fast pace. The methods develop progressively or by fits and starts (just contemplate how drastically PCR or YACs have changed the way DNA studies are tackled). The ideal strategy at a certain point can very quickly become outmoded. This never-ending evolution entails changing course at short intervals to avoid continuing an operation with a now-obsolete technique when some other, five or ten times more efficient, is available. In addition, there is not enough time for methods to stabilize, with the result that nothing is ever fully developed, that procedures are not repeated long enough to settle into a comfortable (and boring) routine. The staff must therefore be able to evaluate new developments, to bring them into use and also to analyze what has happened when the experiment does not yield the expected results: excellent "trouble-shooters" are vital to the success of any major genome laboratory.

Nevertheless, perfecting the human genetic map, constructing contigs of clones spanning tens of megabases and, of course, sequencing large DNA regions do all involve a significant proportion of repetitive work, requiring faultless execution if a ruinous quality-price ratio is to be avoided. The somewhat disorganized improvisation often found in our "academic" laboratories can no longer be tolerated. This organizational shortcoming may be acceptable for small-scale experiments in a university environment in which students are discovering research. However, it has no place in massive programmes utilizing sophisticated robots and – on account of this automation – consuming large amounts of expensive reagents: juste think of the amount of Taq polymerase enzyme used in a US centre like that of Maynard Olson and David Schlessinger at Saint-Louis, where nearly one thousand PCR reactions are carried out daily...

How qualified should the staff be ?

The actual people who do the work, their qualifications and their motivation for these repetitive yet sophisticated tasks are central issues in the management of a genome centre. A very large spectrum of solutions has been tried out. Some managers have adopted an "all-technician" approach, relying heavily on technical assistants trained to perform a specific task. These will, it is hoped, perform reliably and without too much lassitude for several years: as in Tada-Aki Hori's laboratory in Japan, which centralizes cosmid localization by *in situ* hybridization, or at the Généthon in France, which has adopted a similar approach to run some of its sections. Some US Genome Centers, such as the one directed by Rick Myers in San Francisco (chromosome 4) or Glen Evans at the Salk Institute (chromosome 11), function in much the same way. "Technician", in fact, does not have quite the same meaning in the USA as in Europe. In France, such an occupation is usually a lifelong one, often with a stable, civil servant-type position. These professionals do often acquire additional qualifications and, possibly, corresponding increments in rank and salary, but they remain quite distinct from scientists. In the more mobile North-American society, many technicians are in fact students who work in this capacity for two or three years in order to save enough money to go back to university and complete their master's degree or embark on a Ph.D. The result is a high turnover, but also a motivation that routine does not have the time to tarnish.

Other laboratories choose, on the contrary, to employ students and post-docs, thereby creating several problems related to the systematic nature of the work. Though the complete physical mapping of a chromosome is undeniably justified scientifically, it does include a large number of recurring operations. This is true for all recombinant DNA research, that inescapably involves lengthy intervals devoted to preparing DNAs, performing hybridizations and PCR reactions, pouring, running and interpreting gels, etc. The interest in such operations is understandably much greater when the subject is a genetic disease, and when routine is spiced with the hope of finding, for example, "the gene for Huntington's chorea". A partial solution is to assign each researcher two activities: on one hand, participation in a general project, not immediately productive but representing the main thrust of the laboratory's strategy (and often justifying its funding by genome grants); and on the other hand, a personal and more focused topic related to the general theme but likely to prove more interesting and fruitful in terms of publishable material. This tactic may pay off, but there is a risk of drifting off course, the researcher imperceptibly shifting his priority towards the latter project and letting the other activity drop – a very understandable deviation but one to be avoided if the main objective is to be reached.

The trials and tribulations of chromosome-specific libraries

An interesting example is provided by the history of chromosome sorting at the Lawrence Livermore laboratory in California. The Livermore and Los Alamos teams were among the first to succeed in preparing specific libraries from chromosomes isolated using FACS (Fluorescence Activated Cell Sorter) machines. The two centres then shared out the human chromosomes to produce, as a service for the community, a complete set of specific libraries for each of them. Simultaneously, they were pursuing research on perfecting the sorting methods and on applying them to the study of chromosome size polymorphism. Lawrence Livermore chose to assign one particular machine to routine sorting (used to construct libraries) and another one to more research-oriented experiments. Although both activities were supervised by the same scientists, a bias developed resulting in most of the attention being directed to the research activity and the corresponding machine...

It was not long before the quality of the libraries produced (and distributed) by Lawrence Livermore began to suffer: several chromosome preparations were contaminated by *Pseudomonas* bacteria and the ensuing libraries contained a few percent of clones from its DNA. This contamination turned out to be catastrophic for certain users because these libraries were often used to construct sub-libraries of CpG islands, for which various subterfuges are employed in the selection of the clones whose DNA contains CpG-rich sequences corresponding to the "HTF islands", generally associated with genes. As the DNA of *Pseudomonas* contains a high proportion of this dinucleotide, the "HTF island" libraries obtained in this way were largely made up of *Pseudomonas* DNA rather than of human HTF islands... This anecdote illustrates just how dangerous it is to dissociate research and routine; the solution entailed fusing the two again: after first reconditioning the machines, they were assigned alternate shifts of one week for routine sorting and one for research.

An astute blend

I found the best example of integrated systematic study and targeted research in the laboratory of David Ward – leader in the development of non-radioactive probes and persuasive advocate of *in situ* hybridization. His studies on *in situ* hybridization with non-radioactive probes are well known: they have transformed a long and complex technique – valid only in the hands of a few specialists, and relatively inaccurate due to the size of the silver grains signalling the probe – into a reliable and fast method. Using it, a researcher can now localize a hundred or so probes in two or three months, provided that these are cosmids (increasingly the case) and that the laboratory has suitable equipment and expertise. In fact, David Ward's group assigns over a thousand probes per year: cosmids from chromosome-specific libraries, linking clones employed for mapping in pulsed field gels, etc. – in the context of many different projects.

I had imagined a battalion of technicians glued to their microscopes, spending the entire day doing *in situ* hybridization: nothing could be farther from the truth. The laboratory's organization is very instructive. Each of the twelve of so post-docs has an individual project focusing on cloning the gene of a disease, already approximately localized. The work starts with a study of a few dozen probes (generally cosmids) coming from the designated region or from a chromosome-specific library. They are localized by *in situ* hybridization on metaphases and the most strategically placed are then used for high-resolution mapping on interphase nuclei. The area likely to contain the gene is thus narrowed down to several hundred kilobases, in which there are now several probes; this approach is considerably accelerated if the region features a translocation, which is also sought by *in situ* hybridization on interphase nuclei. The post-doc can then switch to YACs and cDNAs in the next step of his research.

With this system, each scientist indulges in a few months of intensive *in situ* hybridization, placing a hundred or so probes as part of his individual project before continuing with other methods. Validity criteria have been clearly defined and a verification process drawn up to confirm the localizations before they are recorded in the database of the laboratory, which thus "produces" a very large number of localizations even though it has only one technician. In other words, David Ward's laboratory employs a very original and efficient working method that appears to resolve the contradiction upsetting many genome projects. However, this solution is only possible because the *in situ* hybridization methods employed here are more advanced than those practised almost everywhere else: a technical advantage that proves decisive. If one or two years were required by Dave Ward's post-docs to localize their hundred probes, or if everyone could localize a hundred cosmids in three months, this imaginative organization would no longer give the laboratory a competitive edge: thus this elegant solution to a major contradiction does not, unfortunately, have general applicability.

Automation and Resource Centres

We have seen in Chapter 9 that instrumentation alone cannot fully resolve the "systematic *versus* focused" dilemma. This does not mean that it must be shunned nor that full use is made today of all its capabilities. However, its implementation is tricky because of the technical, financial and psychological obstacles we have already discussed. One idea is then to divide up the work: the systematic and very codified part could be carried out by a few Service Centres, industrially operated, equipped with machines and well staffed by technicians. The research teams would then use the services of these Centres for the requisite phases of their experiments but would maintain control of their projects and be responsible for devising and performing the most noble, as well as the most tricky, parts of their programmes. In short, a kind of common services pool would be created for the genome.

Common services have a bad reputation in France where they have often performed poorly, and I suspect this is also true in other countries. One specific reason, as far as my country is concerned, is the inflexible organization and staff structure due to an obsession with permanent employment and civil service status, which make reassessments and reorientations very difficult. However, the problems of Resource Centres and other common services ("core services", as they are often called in the USA) extend beyond these administrative questions and are inherent in their function. For one thing, evaluating the future needs of users in quantitative and qualitative terms is not easy, and their forecasts must not be taken too literally as they are often way off the target. These customers can also behave very unreasonably, especially when the service is free: David Schlessinger, whose team generously (and without charge) did a series of screenings by hybridization of the local YAC library for external laboratories in 1988/89, is still bitter about the experience. He has lost count of the number of times that the probes sent in proved to be unusable, and recalls the case of one scientist who, a year after receiving the precious YAC clones extracted not without difficulty from the library, made a second request for them: he had not touched them in the intervening period and had either lost them or left them to dry up in his refrigerator. The fact that the Saint-Louis group switched quickly to PCR screening certainly has to do with the greater reliability of this procedure, even though Hans Lehrach at the ICRF in London would differ on this point. But the main advantage of this method lies in the commitment it requires from the applicant: sequencing his probe, defining and synthesizing the oligonucleotides, checking that they produce a clear and unique band by amplification on genomic DNA... If the applicant has actually completed this work, it is safe to assume he does really need the YAC clone that will eventually be found, and this is reassuring for those who are to undertake the arduous task of screening. Frivolous requests can also be filtered out by having applicants pay a moderate service fee, a measure that is often more effective for its deterrent value than for the income it brings in.

The service Centres also experience difficulty in getting feedback from the users on the adequacy of the libraries, the characteristics of the clones and the results of the investigations in which they play a part. This information is often not usually provided to the "supplier", for whom it would be a valuable asset, if only to assess in detail the quality of his products. In this respect, Hans Lehrach's reference library system (Chapter 5) has the advantage of guaranteeing a minimum amount of feedback since the positive clone(s) detected in the hybridization experiments performed by the outside laboratory can only be supplied when their coordinate(s) on the filter are provided. Thus the central team is automatically informed of which probe has revealed which clone, and this information can be recorded in its database. It is also essential for a service centre to have an exact schedule of conditions and to clearly indicate the services it offers and the applicable circumstances; otherwise, it will soon be subject to interference from parallel circuits. The instructions for using the British HGMP Resource Centre have been clearly set out in G-Nome News and are, no doubt, an example to follow.

Généthon: research or service ?

Généthon, by far the most important heavyweight genome laboratory in France, is the product of a collaboration between CEPH (which provided its expertise) and AFM (the French Muscular Dystrophy Association) which foots approximately 90% of the bill. It has, in fact, a dual purpose: on one hand, several very well set up departments do research on a semi-industrial scale; on the other, Généthon offers services, in particular the provision of ready-made Southern blots and the storage (and "immortalization" by EBV transformation) of blood samples on a large scale.

Research projects include the generation of thousands of CA repeat markers for genetic mapping purposes [13], massive cDNA sequencing, localization of several genetic diseases, and physical mapping by YAC contigs [2, 4]. Each of them employs more than a dozen technicians and is funded at a level of several million US dollars, with the support of an informatics group and an instrumentation development facility. The results and some of the methods developed in these research projects are or will be applicable in more conventional laboratories. The actual service activity involves the production of Southern blots using the now well known "Mark II room" with its 20 robots, and a large DNA and cell bank. The latter activity is particularly welcome, as collection, storage and handling of clinical samples is far from optimal in France (as in some other countries), in part because of relational difficulties between scientists and hospital clinicians. Services provided by Généthon are not free, but their price is much below cost, in effect largely subsidized by AFM as a help to the research community.

Is it worth the money ?

All this work is expensive, very expensive. Equipment costing less than one or two hundred thousand dollars is becoming unusual. Despite the continually dropping prices of hardware, data processing still eats up large sums, and the yardstick used for calcultating the finances of certain centres is now close to $200,000 per person per year – though this includes most of the salaries. These sky-rocketing levels have little in common with the expenses of a standard laboratory, estimated to be two or three times less for a well funded unit concentrating mainly on molecular biology and cell culture. Highly automated and hence consuming large amounts of reagents, open to the outside and "exporting" costly-to-make products, a service activity logically requires more support than conventional research. However making an accurate assessment is very complex and "academic" scientists do not generally have the expertise required. My own impression – very subjective but based on visits to many laboratories and a knowledge of their budgets – is that the quality-price ratio is remarkably variable: depending on the strategy selected, the relevance of the options, how strictly they are implemented and the quality of the centre's management, etc., I would not be surprised to find five or tenfold differences in, for example, the cost per megabase of a YAC contig. In this context, small is not necessarily beautiful: the Généthon whole-genome YAC mapping programme is indeed expensive, but in terms of cost per megabase it may turn out to be a real bargain. All these estimates are obviously difficult, but nevertheless essential since, given the inherent tendency of expansion of the genome programmes, it is vital to spend their large financial allocations in the best possible way.

The emergence of economic issues

The relations between research and industry, and between scientists and money are not usually simple – in fact often antagonistic. The issue of cDNA patenting (discussed in Chapter 7) is a typical example. As a general rule, researchers have a deep-seated aversion for these patents; in their view, there is something almost obscene about the prospect of appropriating part of our genotype based on experiments, that, in the somewhat rash words of Jim Watson, could be carried out by "educated monkeys". These debates have been widely reported in Europe and in France, and have spurred some European decision-makers into premature conclusions: since "the Americans" are taking out patents on the genome, we (French, British...) should protect our own rights by following in their footsteps. These advocates believe we, Europeans, should urgently develop a fast patent process to prevent the sequences produced by our teams from being "stolen" and also ensure this scheme is recognized in the USA. This view seems questionable: in attempting to counter "the Americans", we may in reality be strengthening the hand of the patent lobby, which would then have an easy task of persuading its opponents to accept this approach by arguing that others are doing the same thing.

By spring 1993, the patent issue was still open. The British Medical Research Council – against the general opinion of researchers and the counsel of the senior scientists of its own genome Programme – has registered a patent application for the 1,500 cDNA sequences already produced. In the USA, Jim Watson has been forced out of his position as Director of the NIH Genome Program on account of, among other things, his opposition to these very patents. In favour of this procedure, the "administrators" seem to be gaining the upper hand over the scientists, who are virtually all opposed, while, interestingly, private firms in the USA do not seem over-enthusiastic [1, 9, 11]. The French position, firmly outlined by the Minister of Research in a letter to *Science* [5], is naturally reinforced by the initial rejection of the application by the US patent office, but it is still early days and a reversal of this ruling after an appeal remains perfectly possible...

"Big Science" in biology ?

"Big Science", a term often applied to the genome programmes [6], is something of an overstatement since an annual budget of $200 million remains moderate compared to aerospace programmes or the construction of particle accelerators. Nonetheless, never before has such a structured and coherent project been undertaken in biology: though comparable in scope, the "war against cancer" launched during the Presidency of Richard Nixon or the current research on AIDS both cover a multitude of widely varying studies, most of which taken individually can be classified in the traditional academic category. In contrast, the tasks defined in the genome project are clearly stated: improving the resolution of the genetic map down to two to four centiMorgans, establishing physical maps for whole chromosomes, sequencing large regions of DNA, etc. They can only be accomplished satisfactorily by imposing professional financial management standards and a semi-industrial organization to keep a close watch on project progress and quality control – practices which fundamental research laboratories are not very familiar with. In addition, high flexibility and a capability to reorient experimental strategies must be built in (we have repeatedly stressed the evolving nature of the techniques) – this reassessment must apply both to the organizations and to the persons running them. At times it seems like attempting the impossible – particularly because in all this the motivation of those performing the work must not be overlooked. In this matter, we will have to find a better system of recognition for systematic studies, e.g. by according some merit to the production of information recorded in databases without formal publication (in its usual sense) – as a sort of mini cultural revolution.

Among these emerging questions, many are related to the future of biology in general. Quite evidently this discipline must learn to make better use of instrumentation and to handle stores of information, reagents, clones or other biological objects more efficiently; it cannot postpone for much longer the implementation of a more rigourous organization, at least in some sectors. From this point of view, the full-scale experiment offered by genome programmes is fascinating and it is remarkable to see just how much the issues of human resource management (interest in work, motivation, adaptability, relations between specialists from different disciplines, etc.) occupy the centre of the scene.

REFERENCES

1. Adler RG: Genome research: Fulfilling the public's expectations for knowledge and commercialisation. *Science* 1992 **257:** 908-914
2. Bellanne-Chantelot C, Lacroix B, Ougen P, Billault A, Beaufils S et al: Mapping the whole human genome by fingerprinting yeast artificial chromosomes. *Cell* 1992 **70:** 1059-1068
3. Bodmer WF: Two cheers for Genome sequencing. *The Scientist* October 20 1986: 11-12
4. Chumakov I, Rigault P, Guillou S, Ougen P, Billaut A et al: A continuum of overlapping clones spanning the entire human chromosome 21q. *Nature* 1992 **359:** 380-387
5. CurienH: The Human Genome project and patents. *Science* 1991 **254:** 1710
6. Davis BD: Human Genome project: is "Big Science" bad for Biology? Yes, it bureaucratizes, politicizes research. *The Scientist* 1990 **4:** 13-15

7. Davis BD and Colleagues: The Human Genome and other initiatives. *Science* 1990 **249:** 342-343
8. Dulbecco R: A turning point in cancer research sequencing the Human Genome. *Science* 1986 **231:** 1055-1056
9. Eisenberg RS: Genes, patents, and product development. *Science* 1992 **257:** 903-908
10. Gilbert W: Genome sequencing creating a new Biology for the twenty-first century. *Iss Sci Tech* 1987 **3:** 26-35
11. Kiley TD: Patents on random complementary DNA fragments? *Science* 1992 **257:** 915-918
12. Palca J: Human Genome sequencing plan wins unanimous approval in US. *Nature* 1987 **326:** 429
13. Weissenbach F, Gyapay G, Dib C, Vignal A, Morissette F et al: A second generation linkage map of the human genome based on highly informative microsatellite loci. *Nature* 1992 **359:** 794-801

14
The ugly ethician

Beware of ethics: trespassers will be prosecuted !

Very early on, when this book was but a figment of my imagination – the survey itself being still at the project stage – it was examined by an INSERM Committee, which suggested the appointment of a co-author to cover ethical issues raised by genome programs. Admittedly my knowledge in this field is rather shallow: I am not an established ethics expert, nor a priest, a lawyer nor even MD. Nonetheless, I will allow myself the liberty of treating this sensitive subject from a personal angle, drawing on my perceptions during this study. Ethical issues as such have been thoroughly aired – though by no means resolved – in many papers and publications [1, 4]: a bibliography compiled by Michael Yesley for DOE [4] contains more than 2,600 references... Thus I see no real need for me to bring my own feeble light to bear on the subject. Neither does it seem particularly relevant to add to this book an "ethical issues" annex, written by a specialist, that would summarize thoughts developed more thoroughly elsewhere. I have thus preferred to discuss what I saw and heard during my visits to the laboratories, and to analyze how they experience ethical issues as well as interact and communicate with the society to which they belong.

Fallout from genome programmes

A quick overview of the new possibilities opened up by genome research and of the risks associated with "predictive medicine" is in order before discussing how they are perceived in laboratories.

Poorly controlled repercussions

The impact of the genome programmes on society as a whole is far from insignificant. The new knowledge thus gained leads to the elimination of embryos through prenatal diagnostics and pregnancy termination. It enables prediction of the future for as yet asymptomatic individuals – a catastrophic forecast sometimes, as in the case of Huntington's chorea. It also opens the possibility of cataloguing humans according to the "quality" of their genotype, a classification that naturally opens the door to discrimination, to exclusion of bad health risks from private insurance systems and to all possible sorts of ostracism.

Without comparable progress in therapy (which generally advances much more slowly), this knowledge creates a widening gap between our new diagnostic capabilities and the narrow path of treatment. The intense anguish aroused by some investigations makes us wonder, for example, whether it is desirable to perform diagnostics for late-onset diseases having no known remedy.

Last of all, this knowledge on the genome creates the risk of an intervention, of "genetic manipulation" *proprio sensu*. For man, this danger is more fantasy than fact since all the performed or planned gene therapies are somatic and involve grafting a functional gene solely in the cells of an organ without modifying either the germline or the genotype transmitted to descendants. However, such work is indeed performed on animals. Laboratory mice are being modified to transform them into models for inheritable diseases. The production of a genetically sick mouse in which both copies of the CFTR gene have been inactivated and which presents some of the symptoms of cystic fibrosis [3] is a recent landmark in this respect. Similarly, foreign genes are introduced into livestock to improve their productivity or to obtain milk containing molecules of pharmaceutical value. Understandably, some people question our right to so manipulate the genetic makeup of animals, and fear that similar operations may one day be practised on man.

Health care systems: private enterprise or state control ?

These issues cannot be discussed without considering another parameter, the nature of the health care system. It has a strong influence on the speed at which technical advances are put into medical practice. It also determines which category of the population will be the first to benefit from them. Few regulatory barriers exist in free market systems, such as in the USA, where private laboratories perform karotypes and genetic analyses on request. This is quite expensive, but any person who can afford it may have the sex of a foetus diagnosed at 2 months, may find out whether she carries the delta F 508 mutation or (for a man) if he is indeed the biological father of his child. In state-run or highly supervised health care systems, as they exist in France or Great Britain, this is virtually impossible:

such examinations are under regulatory control and are solely authorized within the framework of a public hospital (or of a legal investigation). Consequently some technical advances are transferred into medical practice more slowly in these countries than on the other side of the Atlantic; it may be said that in the USA users get better service, while researchers enjoy more direct access to the commercial application of their discoveries.

However, this is a double-edged advantage. Premature commercialisation and biased interpretation of results can thrive in this context, as evidenced by the controversy surrounding DNA-based identification tests, the famous "genetic fingerprints". Based on minisatellite analysis, this technique was introduced in Great Britain as early as 1987. First applied to immigration cases (a family kinship had to be proved for eligibility to enter British territory), it was then used to identify suspects in criminal cases. All this took place without arousing strong objections: the method had been implemented under the close supervision of its inventors, and with copious precautions. In the USA, in contrast, several companies created for the purpose rushed headlong into this market and the legal system accepted their word as gospel truth – at least in the beginning, until it was realized that some of them did not really master these methods, that the validity of their results was more than suspect and that even the statistics presented in their conclusions were biased. To cope with the ensuing outcry – which nearly discredited this technique, very efficient when properly applied – eminent scientists, such as Eric Lander, had to be brought in to specify quality criteria, and the National Academy of Sciences ended up designing strict guidelines.

Speed, therefore, does not always pay off, notably in fields as sensitive as human health care or forensic medicine. Moreover, in the USA, it is associated with a two-tier health system: innovations are available for those who can afford them, i.e. who have good medical insurance coverage. And this brings us to a touchy question: should the insurer have the right to know the genotype of a potential customer wanting to take out a policy. This problem is likely to become rapidly insoluble to the point that numerous US authorities believe society will have to take charge of health risks, in complete contradiction with the country's underlying free market ideology.

Real-life experience of ethics

These advances and these difficulties result from the research conducted in the human genetics laboratories; how are these repercussions seen by those who are at their origin ? I will attempt to address this question through my personal experience.

Possibly damaging catch-words

First let us lend an ear and listen to the language used in laboratories: it is likely to shock the unacquainted visitor. Researchers talking among themselves will refer to "human material", which smacks of the Third Reich's "Menschmaterial" but in fact, refers simply to samples of human blood or cell lines. They may also mention the "cystic fibrosis gene" instead of using the more correct (but more laborious) "the gene which, when defective, is the cause of cystic fibrosis". Even the expression "defective gene" is indeed full of presumed value judgements (good genes, bad genes) and probably the label "non-functional" would be more appropriate. Scientists will also discuss, sometimes in writing, "the alcoholism gene" whereas the role of heredity in this behaviour is still very debatable. Another often-used term to denote a family tree is "pedigree" (certainly a valid word but with a connotation of "breeding" for the French general public and, I suspect, for others as well); as for the expression "a perfect case", it is commonly employed to indicate a child with a particularly characteristic form of a dreadful disease. This is not too serious a matter and researchers are probably neither more or less human than the rest of mankind. Nevertheless, the use of these terms, which is open to misinterpretation, does perhaps reflect an excessive alienation from the consequences of their work.

The benefits of contact with the patients

Attitudes do, in fact, vary widely. In those laboratories close to clinical genetics, often located in hospitals, the scientists interact with doctors or are themselves MDs, and they are fully aware of the individual and family tragedy associated with muscular dystrophy or cystic fibrosis. As a result, they are quite conversant with the issues that modern genetics raises for society. On the other hand, many researchers in major centres focusing on the "heavy" genome activities tend, in my experience, to be very engrossed in their YACs, contigs or kilobases and little inclined to meditating on the consequences of their work. When they do, they are sometimes very naive, like the researcher who dreamt of constructing a "25th chromosome" carrying large gene regions to provide a definite cure for certain inherited afflictions – without thinking too much about the future of the new human race he would thus be creating.

Disease advocacy groups play an important part in bringing scientists into contact with the day-to-day reality of a disease as well as with the needs of the patient and family. A good illustration of this is provided by the meetings organized by the Fragile X Foundation in the USA: the scientific content is excellent, specialists give papers and participate in learned discussions, but attention is also paid to schooling and behavioural problems and to psychological support for parents with afflicted children. The researchers who are there to outline the advances made also meet patients' parents and are confronted with their outlook. For these men and women, the name of the laboratory that won the race to clone the FMR 1 gene (see Chapter 6) is completely irrelevant; what is vitally important for them is speedy translation of this advance into progress in diagnostics and therapy. Similarly, the Telethon campaigns in France are the occasion for "open house" days in many laboratories subsidized by AFM: researchers meet, sometimes for the first time, with people suffering from the disease they have been working on for months or years, and the outcome is on occasion surprising.

A cut for ethics

The senior officials in genome research have in general acknowledged the need to tackle the issue of ethics. In the USA, 5% of the funding provided by NIH and DOE is earmarked for the Ethical Legal Social Issues (ELSI) program, a large sum amounting to millions of dollars given the overall genome budget of about $200 million. Eric Juengst for NIH, and Michael Yesley as an advisor to DOE are in charge of these programmes. They are respectively a philosopher specialized in ethics and a jurist who studied the protection of persons involved in medical and sociological experimentation. A quick glance through their projects * reveals many interdisciplinary meetings but also research on the quality control of genetic analyses, the attitudes of doctors regarding these new methods, the psychological consequences of a positive diagnosis for Huntington's chorea and the potential socio-economic impact of widespread testing for cystic fibrosis carriers. Another of their concerns is the production of audio-visual bulletins for the general public.

The European project also allocates sizeable funds, one million Ecus (about 7% of the total budget) to the study of ethical issues. The topics to be addressed are largely similar to those selected in the USA, except that the variety of situations among EEC member countries makes an enquiry and inventory phase essential. For instance, just consider the different notions of filiation existing in two countries as close as France and Germany: in the former, the social aspect is privileged and steps have recently been taken to prevent the genetic fingerprinting technique from being employed other than in a legal action. On the contrary, across the Rhine, the biological filiation is primordial and the right of the child to know his biological father is strongly emphasized. It would be easy to multiply the number of these examples and it is understandable that this diversity must be taken into account before attempting to define a common position.

---

* *Human Genome News* 1991 **2:** 10-11

The melting pot of science

Marked differences in laboratory moods and attitudes towards ethics could be expected according to the country, given the variety of national cultures. Yet this diversity is limited from the start, since there are not many nations affluent enough to finance a major genome programme: as we have seen, the map of this research is quite similar to that of the rich countries, the USA, Japan and Europe essentially. Nevertheless, each of these nations has its own linguistic and cultural flavour; but these have little apparent impact in the laboratories: the world of scientists is much more unified than the societies they are living in. The universal use of English, the frequent trips, the sabbatical or post-doctoral stays abroad, the high proportion of foreigners in laboratories are all factors tending to enforce cultural uniformity based on the Anglo-Saxon model, the dominant one. Thus national culture does not have a decisive influence on the laboratory environment, even in Japan – as far as a "gaijin" (foreign barbarian) can judge, for the Japanese have a disconcerting knack of appearing westernized and then, on occasion, turning out to be well and truly Japanese. Everywhere in the world the reading matter consists of *Nature*, *Science*, *Cell* and *PNAS*, the equipment used is mostly of North-American origin and the issues of ethics and the effects on society are debated in a similar way – that is, they do not arouse very much interest...

Society's outlook on research

Real differences do, however, exist in how society perceives research, and this determines the tone of the relations between these two worlds. This topic will be discussed using two very dissimilar examples: France and Germany.

France: an almost suspect indulgence

The exceptional nature of the situation in France is often not realized. Science, in that country, is undeniably crowned with an aura of prestige, and the public image of the researcher is very positive. The ideological discords expressed at the time of the May 1968 demonstrations – later to crystallize around opposition to "genetic manipulations" (the name given to recombinant DNA research by its opponents in the seventies) – have remained confined to a small circle; the recent growth of the ecologists as an influent political movement has not given rise to overall "anti-science" feeling. A positive attitude to science is quite commonplace in France, doubtless the only developed country in which nuclear power is so well accepted and where advances in biology arouse little fear. Is this due to a traditional esteem for the intellectual, to some well conducted industrial policies (e.g. the way Électricité de France has run nuclear power programmes, which now provide two thirds of the electricity) or, on the contrary, is it a sign of resignation and abdication in the face of authority ? Regardless of the reasons, France is one of the countries where obstacles to the practice of research are the least frequent. This attitude is not to be interpreted as indifference; on the contrary, there is a lively interest in science, as attested by the existence of numerous science magazines. In addition to the very serious-minded *La Recherche* or *Pour La Science* (the French edition of *Scientific American*), there are several popular magazines such as *Science et Vie* (with a circulation of 300,000), *Sciences et Avenir*, *Ça m'intéresse*, etc., that constitute a strong affirmation of public interest in research. Yet another sign of this mobilization is the success of the TV campaigns organized by AFM.

French scientists thus have the good fortune of operating in a society which values them: a privileged and clearly very advantageous situation that, however, does imply a few responsibilities. French public opinion expects a great deal from its researchers. Above all, it whishes to see quick progress in medicine, and may be diasappointed insofar as it was led to believe that the discovery of a disease gene would be rapidly translated into improved treatment. It also wants to have the new discoveries explained, as the success of the popular scientific magazines demonstrates. And finally the French public does not understand how scientists can be indifferent to the possible repercussions of their work and expects them to take part in curbing any harmful tendencies. Such a symbiosis between science and society can only be maintained if these expectations are fulfilled - or, in certain cases, if it is clearly explained why they are unrealistic.

Recent events must be mentioned here. The still ongoing row about HIV contamination of hemophiliacs is, I feel, a paradoxical reflection of this high regard for medical scientists, accompanied by high expectations from them. The delay between recognition of HIV transmission by blood products and the provision of safe, heat-treated clotting factors - disastrous and unjustifiable as it is - has been similar in France and in other European countries, yet only in this country has it developed into a full-blown scandal threatening a former prime minister. Apart from other factors, it seems that this indicates how crestfallen public opinion is at discovering that scientific authorities can be unreliable or incapable of decisive action...

In Germany: a witch hunt !

The situation across the Rhine is very different. As we have already mentioned (Chapter 11), the Germans generally view biological research with a great deal of suspicion. This stems from a combination of reasons, the first being a strong ecological movement with a marked anti-science attitude. Based on a healthy reaction against the power of the experts, sometimes guilty of misusing their privileged position, this stand ends up rejecting all scientific reasoning and, in consequence, all rationality in favour of vague and magic concepts such as "natural", "biological", etc. In this setting, the relatively recent technology of recombinant DNA has taken on such sinister overtones that very stringent legislation has been passed, causing several major firms to move their biological research and production activities outside the country. And just to make things worse, the German human genetics community heavily compromised itself in the past with the Nazis, making the climate today frankly hostile to work on the genome; German colleagues talk of aggressiveness, ostracism and sometimes even threats – or, in extreme cases, bomb attacks.

Given the prevailing mood, discussions with the opponents of genome research are taxing. I have met scientists who have tried to play the game and attended public meetings with open debates. They had the impression that opinions were already fixed and they themselves, being specialists, were *ipso facto* suspect from the start. Confronted with this opposition, the temptation is strong to stay closeted in the laboratory, to change fields and work on a less contested topic – or to relocate elsewhere and continue the work in calmer waters. Thus it is clearly very difficult in Germany for the research scientists and the public to communicate with each other. This has disturbing consequences both for the scientists, whose work is hampered, and for the public, for whom the availability of certain medical breakthroughs will be delayed.

The duty of communication

What can be done to avoid such a deadlocked situation? It seems that the duty (as well as self-interest) of researchers lies in communicating with the entire society, to present a rational view of their work to allow the derivation of suitable mechanisms for controlling the resulting applications. However, life is never simple and this communication does not materialize easily.

"Just let us get on with our work!"

This is quite a common reaction of researchers when solicited by the public or the press. Negative attitudes to communication are widespread, ranging from the head of a US laboratory who refused to meet me (let me stress that this was a unique occurrence) on the grounds that "enough has been said on the genome, it's time to get back to work", to the young scientist who "doesn't have the time" to talk to a high-school teacher. These attitudes are understandable: communicating does indeed take time, a lot of time. People seeking information – especially journalists – do not always seem to realize this. Moreover, communicating with the public automatically implies a simplified and schematic description. The rules in this case differ markedly from those of the "primary" publication, a fact not understood by everyone. An interview with a reporter, published in a newspaper or a magazine, generally prompts critical comments from colleagues – ironical at the best, outraged at the worst – because some name has not been mentioned or because a result has been presented without all the careful qualifications normally included in a research paper. Furthermore, journalists do on occasion make mistakes and, even when the article has been written by a scientist, the all-powerful editing board sometimes assumes the right to change the text or to add a title that is catchy but ill-suited to the content. In short, "informing" is not very well regarded in the research community. Many choose to stay in the background and those devoting time to this activity are often considered as second-rate researchers, who have made up for their relative failure by one of those "arabesques" so dear to the immortal Peter [2] and who have cloistered themselves in what one of my colleagues once called "the tertiary sector of research"... a term that strongly expressed his lack of esteem for this less than noble occupation.

A strong demand

Yet the genome Programme has undeniably created much interest both within and beyond the scientific community. This comes as no surprise: its ultimate objectives are understandable, surely more so than research in physics on the intermediate boson particle or than the launch of a space telescope. There is, in fact, something miraculous about the future prospect of being able to fully decipher the messages hidden in our genetic make-up. In addition, the connection with health care is immediately obvious, and, in France, the Telethons organized by AFM have proved very effective in informing and influencing public opinion. These TV programs – now an institution – have popularized (in the positive sense of the word) the concept of genetic diseases and have explained the connection between fundamental biological research and therapeutical progress – even if a few misconceptions were allowed to propagate as to its probable rapidity. Yet at the same time research on the genome creates uneasiness: it is felt to touch something very fundamental, deep down in our innermost recesses, to constitute a dimly perceived menace. It is also understood that it will create new possibilities for both preventive and curative medicine, whose correct application may quite rightly be questioned.

The members of the public are therefore asking for information. A justifiable request: after all, they are paying for the research (via income tax or donations), and it is their lives that will be favourably or unfavourably affected by the impact of this work. They constitute the society that will have to set down limits and safety measures and to define the "guidelines" that technological progress will make indispensable. It is thus quite normal that the public should be kept informed of how this program is being conducted, should be aware of its repercussions and of the ethical issues, most of which are genuine problems requiring solutions.

The role of research scientists

The researcher's place in this information process is not obvious. Not a journalist, he has neither the inclination, time, nor competence to communicate directly with the public; neither is he in a position to make decisions on the problems of society, even though he may be solicited to give advice on the validity of a given technical solution. In addition, he is expected to concentrate on his research topic, which leaves little time for extracurricular activities. Nevertheless in my view, he must devote some of his time to these peripheral activities. Indeed, in France, providing scientific and technical information is an explicit part of the duties of INSERM and CNRS employees. To save time, the scientist is pushed into screening his correspondents, only retaining those who have a basic knowledge of the subject: there is no more trying activity than having to spend an hour over the telephone teaching modern biology to successive reporters, all virtually illiterate on the subject ! One of the secrets for the success of scientific columnists such as Gina Kolata (previously with *Science*, now at the *New York Times*) or Leslie Roberts (who frequently publishes articles or news on the genome in *Science*), is precisely their in-depth knowledge of the research sector they are covering, which enormously facilitates contacts with scientists.

It is also advisable to set up relays, which can be people: journalists need not always talk to the "boss", young scientists may be delighted to have the chance and may find food for thought in the process. Articles in the popular press also have a part to play as does participation in audio-visual productions. INSERM has, it should be mentioned, innovated in this matter by producing seven-minute TV clips focusing on subjects such as stress, AIDS or gene therapy. Their originality lies in the participation of an "INSERM Youth Club", a group of high-school students who plan the scenario with the help of a researcher. Neither should the teachers, who are trained to pass on knowledge, be left out. Those having the widest contact with the public are the secondary-school professors of biology or natural sciences. Most of them graduated before the DNA era: the recombinant DNA revolution occured barely fifteen years ago and, in the intervening period, our knowledge probably progressed more than in the last 150 years. In this area there is an overwhelming and widespread need for training, for which novel solutions must be found. In the USA, a major retraining program has been developed within the framework of the National Science Foundation (NSF), and implemented, among others, by Leroy Hood and colleagues at Caltech. It is backed by substantial resources and involves the active participation of numerous researchers from the laboratory.

There are thus many possible ways for the researcher and society to communicate with each other; it is feasible, with a little forethought, to do this without jeopardizing one's research and, in doing so, to see one's work from a new angle.

REFERENCES

1. Murray T: Ethical issues in human genome research. *FASEB J* 1991 **5:** 55-60
2. Peter LJ, Hull R: *The Peter principle.* William Morrow, New York, 1969
3. Snouwaert JN, Brigman KK, Latour AM, Malouf NN, Boucher RC et al: An animal model for cystic fibrosis made by gene targeting. *Science* 1992 **257:** 1083-1088
4. Yesley SM: Bibliography: Ethical, legal and social implications of the Human Genome project (May 1992) DOE, Office of Energy Research, Washington DC 20585 (document DOE/ER-0543T)

Postface

May I conclude this book with a few personal considerations ? It is the fruit of a sabbatical year which I had organized so that I would have the widest possible spectrum of options on my return: I had not only left the directorship of my institute but also closed down my team and my associates had gone abroad for post-doctoral training. To some extent therefore I had burnt my boats, motivated as I was by two objectives: a wish to be as free as possible during a survey year that I expected to be intensive, and an uncertainty as to my future activities. These would not necessarily include, I felt, a direct involvement in research given my motivation for communication.

During this itinerant year, I was able to test myself as a full-time communicator. My functions during these twelve months amounted to travelling, visiting, discussing and writing. This was an enriching experience due to the variety of colleagues, institutions and cultures whose way of life I was able to sample. Throughout my interviews, however, I became aware of how important it was for them to be talking to a fellow scientist – rather than to a journalist – and how my own assessment of the laboratories was sharpened by my being an experimentalist. Very soon I felt the urge to be involved, to try out assumptions and to devise experiments. Six months after my departure, my strong interest in communication was confirmed – but in communication *as a scientist*. To continue being a real scientist meant becoming involved again in research, preferably in the framework of a small team allowing direct contact with experiments. I have followed this course of action, given careful consideration to topics, methods and possibilities, and formed a new group in my original laboratory. I am thus back, if not to the bench, at least to a direct contact with, and responsibility in, research. This is what I wish to do. However I shall certainly not give up communicating on research – as a scientist !

Abbreviations

AAtDB: An *Arabidopsis thaliana* Data Base (patterned after ACeDB).

ACeDB: A *Caenorhabditis elegans* Data Base.

AFM: Atomic Force Microscope.

AFM: Association Française contre les Myopathies (French Muscular Dystrophy Association).

ARC: Association pour la Recherche sur le Cancer (one of the two major French cancer research associations).

ATCC: American Type Culture Collection.

BAC: Bacterial Artificial Chromosome.

CEA: Commissariat à l'Énergie Atomique (the French DOE).

CEPH: Centre d'Étude du Polymorphisme Humain (Centre for the Study of Human Polymorphism).

cDNA: complementary DNA.

CNRS: Centre National de la Recherche Scientifique (National Centre for Scientific Research, somewhat similar to NSF).

DOE: Department of Energy (USA).

EEC: European Economic Community.

EST: Expressed Sequence Tag.

EMBL: European Molecular Biology Laboratory, in Heidelberg (Germany).

EMBO: European Molecular Biology Organization.

FACS: Fluorescence-Activated Cell Sorter.

GDB: Genome Data Base (Baltimore, USA).

HGMP: Human Genome Mapping Project (the British genome Programme).

HGMW: Human Gene Mapping Workshops.

HHMI: Howard Hughes Medical Institute (USA).

HUGO: Human Genome Organization.

ICI: Imperial Chemical Industries (Great Britain).

ICRF: Imperial Cancer Research Fund (Great Britain).

IGD: Integrated Genome Database (Heidelberg, Germany).

INSERM: Institut National de la Santé et de la Recherche Médicale (the French NIH).

MAC: Mammalian Artificial Chromosome.

MRC: Medical Research Council (Great Britain).

NIH: National Institutes of Health (USA).

NRC: National Research Council (USA).

PCR: Polymerase Chain Reaction.

RFLP: Restriction Fragment Length Polymorphism.

SICM: Scanning Ion-Conductance Microscope.

STA: Science and Technology Agency (Japan).

STM: Scanning Tunnelling Microscope.

STS: Sequence Tagged Sites.

YAC: Yeast Artificial Chromosome.

Subject index

Numbers indicated in bold type refer to pages where the items are detailed.

A

AFM, see French muscular dystrophy association or Atomic force microscope
Alcoholism and genetics, 170
Alu, 45
Alzheimer, 146
American type culture collection (ATCC), 56
ATCC, see American type culture collection
Atomic force microscope (AFM), 90

B

Bacterial artificial chromosome (BAC), 47
Bacteriophage P1, 47
Belgium, 141
Big science, 163

C

CA repeats, see Microsatellites
Caenorhabditis elegans (Nematode), **81**, 128
Canada, 83
Candidate genes, 65, 84
cDNA, **11**, 84, **114**
— clones, 86, 162
— patents, 87, 135, 163
CentiMorgan, 8, **28**
CEPH (Center for the study of human polymorphism), 13, 108, 131, 133, 161
Charcot-Marie-Tooth disease, 145
Chiba laboratory, 120
Chimeric clones, 41, **42**
Chromosome 11, **129**
— 19, **54**
— —, physical map, 83, 98
— 21, genetic map, 30, 146
— X, 35, 83
— editors, 100
— painting, 45
— size polymorphism, 158
— sorting, 158
Cold spring harbor genome mapping, 117
Computerized laboratory notebook, 98
Contigs of DNA clones, 8, 52, 98
Cosmids, 46, 54, 128
— contigs, 131
— libraries, 98
Cystic fibrosis, 2, 45, 53, 65, 146

D

Denmark, 141
Department of energy (DOE), 12, 54, 91, 92, 98, 100, **112**, 151, 171
Deutsche Krebsforschungs Zenter (DKFZ), 139
DNA cloning, 35
— sequencing, 155
DOE, see Department of energy
Down syndrome, 146
Duchenne muscular dystrophy, 2, 53, 146
Dystrophin gene, 53

E

EEC, 15, 102, 133, 142, 151, 171
ELSI, see Ethical legal social issues
EMBL, see European molecular biology laboratory
EMBO, see European molecular biology organisation

Escherichia coli, 8, 10, 51
ESF, see European science foundation
EST, see Expressed sequence tags
Ethical legal social issues (ELSI), 171
EUROGEM (European gene mapping), 141
Europe, 137, **141**
European data resource for human genome analysis, 139
— molecular biology laboratory (EMBL), 100, 139, 148
— — — organisation (EMBO), **147**, 148
— science foundation (ESF), bibliometric study, 14
Exon trapping, 65, 131
Expressed sequence tags (EST), **57**

F

Filiation, 171
Fingerprints, 129
Finland, 140
FMR-1 (fragile mental X retardation), gene, 67, 72, 140
Fragile site of the X chromosome, **66**
Fragile X-linked mental retardation, 66
— chromosome, **63**, 131, 140
— foundation, 170
France, **131**, 172
French muscular dystrophy association (AFM), 15, 108, 126, 132, 161, 171

G

G-Nome News, 126, 161
G-String, 126
GDB, see Genome data base
Genatlas, 100, 126
GenBank, 100
Généthon, 28, 55, 108, 126, 161
Genetic manipulations, 172
— fingerprints, 169
Genexpress, 87, 115
Genome centers, **112**, 126
— data base (GDB), 100, 101, 115, 121, 140, 149, 151
— and ethics, **167**
— mapping and sequencing, 147
Germany, 139, **173**
Great Britain, 2, 87, 115, **125**
GREG, 136
GT repeats, see Microsatellites

H

HGMP, see Human genome mapping project
HGMW, see Human gene mapping workshops
Homologous recombination (YACs), 45
Howard Hughes medical institute (HHMI), 12, 100, 148, 151
HUGA, 105, 109
HUGO, see Human genome organisation
Human frontier, 13, 117
— gene mapping workshops (HGMW), 100, 136, 140, 149, 150
— genome mapping project (HGMP), 125, 161
— — organisation (HUGO), 12, 15, 88, 111, 145, **147**
— genome programmes, 2, **65**
Huntington's chorea, 53, 146, 168, 171
Hypermethylation of DNA, 71

I

ICRF, see Imperial cancer research fund
Immortalization, 162
Imperial cancer research fund (ICRF), 13, 58, 70, 126, 127
In situ hybridization, 33, 159, 160
Informatics and genome, **97**
Interphase nuclei, 159
Irradiation hybrids, 131
Italy, 2, **137**

J

Japan, 2, 105, **116**
— human frontier programme, 117

K

Kallman syndrome, 84, 131
Kennedy disease, 73

L

LABIMAP (Eureka programme), 109, 134
Libraries, chromosome-specific, **99**
— cosmid, 58
— reference, **58**, 127
— YAC, 8, 36, **40**, 58, 98, 133
Linking clones, 159

M

MAC, see Mammalian artificial chromosomes
Mammalian artificial chromosomes (MAC), 46
Maps, genetic, 6, 7, **8**, 58, 113
— integrated, **51**
— physical, 6, 7, **8**, **51**, 58, 127
— whole-chromosome, 33
— whole-genome, 17, **33**
Mark II, 29, **108**, 162
Medical research council (MRC), 88, 126, 130, 163
Megasequencing, 80
Microdissection, 67
Microsatellites (CA repeats, GT repeats), **24**, 149
— and PCR, 57
Minichromosomes (double minute), 48
Minisatellites (VNTR), 23
Monbusho, genome project, **116**
MRC, see Medical research council

N

National institutes of health (NIH), 12, 19, 23, 87, 100, **112**, 151, 156, 163, 171
— research council (NRC), 19
— science foundation (NSF), 176
Nematode (*Caenorhabditis elegans*), 81, 128
Netherlands, **140**
Northern blot, **21**
NIH, see National institutes of health
NRC, see National research council
NSF, see National science foundation

O

Oligonucleotide ligation assay (OLA), 141
OMIM (On line mendelian inheritance in man), 101

P

Patents, **87**, 163
PCR (polymerase chain reaction), 24, 25, **27**, 113

Polymerase chain reaction (PCR), 24, 25, **27**, 113
Polymorphism, **23**
—, restriction, 18
Positional cloning, see Reverse genetics
«Predictive Medicine" EEC project, 15, 141
Predictive medicine, 167
Probes, 149
—, internal and end, 43
Pseudomonas, 159
Pulsed field gel electrophoresis, 52

R

Resource centre, **126**, 160
Restriction fragment length polymorphism (RFLP), **17**, 23, 149
Reverse genetics (also called positional cloning), **1**, 19, **63**
RFLP, see Restriction fragment length polymorphism
Rice genome project, 121
Riken life science center (Tsukuba), 118
Robotics and genome, 104

S

Scanning ion-conductance microscope (SICM), 90
— tunnelling microscope (STM), **90**
Science and technology agency (STA) (Japan), 118
Scandinavia, **140**
Sequence tagged sites (STS), **55**, **57**, 113, 128, 149
Sequencer, 79
— personal, 122
Sequencing, **11**, 155
— DNA, 77
— —, automatic, 105
— cDNA, 83
— *Caenorhabditis elegans*, 81
— capillary electrophoresis, 11
— Yeast chromosome III, 82, 83
—, STM, 11, 89, **90**
—, whole-genome, 84
—, HUGA system, 106
— meeting, 117

SICM, see Scanning ion-conductance microscope
"Signature", 85, 98, 151
Somatic hybrids, 70
Southern blot, **21**
Spain, 141
Sperm typing, 27
STA, see Science and technology agency
Stealth vector, 47
Steinert muscular dystrophy, 73
STM, see Scanning tunnelling microscope
STS, see Sequence tagged sites
Sweden, 140
Switzerland, 139

T

Telethon, 132, 171, 175
Translocations, **68**
Tunnelling microscopy (STM), **90**

U

UNESCO, 146
United States of America, 2, 5, 22, 88, **112**, **131**, 146, 151

V

VNTR (variable number of tandem repeat), see Minisatellites

Y

YAC (yeast artificial chromosomes), 8, **36**, 45, 113, 131
— libraries, see Libraries
— clones, 39, 41
— homologous recombination, 45
— transfer, 46
— and fragile X, 68
Yeast, **82**
Yeast artificial chromosomes, see YAC
Yeast chromosome III, **82**

Author index

Adler, Reid, 87
Albertsen, Hans, 41
Allshire, David, 129
Anand, Rakesh, 41, 114
Ansorge, Wilhelm, 139
Auffray, Charles, 86, 115, 134
Avery, Oswald, 1

Baldeschwieler, John, 93
Balhorn, Rod, 92
Ballabio, Andrea, 137
Bentley, David, 52
Bird, Adrian, 130
Bodmer, Walter, 13, 126, 148, 152
Botstein, David, 82
Bourke, Frederick, 77
Brenner, Sydney, 13, 86, 87, 126, 127, 128, 147
Burke, David, 34, 36, 71

Cantor, Charles, 34, 112, 148, 152
Carle, Georges, 34, 36, 71
Carrano, Anthony, 54, 83, 98, 112, 137
Caskey, Tom, 112, 113, 152
Cavalli-Sforza, Luca, 138, 139
Church, George, 11, 80, 112
Cohen, Daniel, 41, 45, 108, 134, 151
Collins, Francis, 112, 116
Cooke, Howard, 129
Coulson, Alan, 81, 128
Crick, Francis, 1

Dausset, Jean, 13, 131, 133, 148
Davies, Kay, 68, 69, 72
Donis-Keller, Helen, 8, 9, 17, 19, 23
Dulbecco, Renato, 137

Emanuel, Beverly, 112
Endo, Isai, 105
Evans, Elizabeth, 152
Evans, Glen, 112, 157

Frézal, Jean, 100

Gesteland, Ray, 112
Gibson, Keith, 126

Hanoune, Jacques, 136
Hastie, Nick, 129
Honjo, Tasaku, 121
Hood, Leroy, 80, 141, 176
Hori, Tada-Aki, 157
Horsthemke, Bernard, 68

Ikawa, Yoshi, 120
Imai, Takashi, 120
Isono, Katsumi, 120

Jacobson, Bruce, 89
Jeffreys, Alec, 23
Juengst, Eric, 171

Kanehisa, Minoru, 120
Keller, Richard, 89
Klepsch, Andreas, 138
Kohara, Yuji, 8
Kolata, Gina, 176
Kourilsky, Philippe, 135

Laird, Charles, 71
Landegren, Ulf, 141
Lander, Eric, 112, 169

Lazar, Philippe, 135
Lehrach, Hans, 45, 58, 127, 128, 151, 161
Le Paslier, Denis, 41
Little, Peter, 129

Mach, Bernard, 139
Mandel, Jean-Louis, 69, 71, 73, 131
Matsubara, Kenichi, 115, 116, 118, 119, 120, 121, 148
Matsuda, Fumihiko, 119, 121
McKuzick, Victor, 96, 147, 148
McLeod, Charles, 1
Mirzabekov, Andrei, 138, 148
Monaco, Tony, 41, 114
Moysis, Bob, 112
Muller-Hill, Benno, 139
Munnich, Arnold, 108, 131
Myers, Rick, 112, 157

Nakamura, Yusuke, 120
Nelson, David, 71

Okayama, Hiroto, 120
Olson, Maynard, 34, 41, 55, 56, 71, 151, 157
Ommen, Gert-Jan van, 138, 140
Oostra, Ben, 71, 140

Pearson, Peter, 100, 115, 140, 149
Petit, Christine, 84, 131
Petterson, Ulf, 141
Porteous, David, 129

Rine, Jasper, 112
Roberts, Leslie, 176
Rysavy, Francis, 126

Schlessinger, David, 41, 69, 71, 112, 113, 137, 157
Schwartz, Charles, 34
Sibson, Ross, 126
Siniscalco, Marcello, 137, 138
Sinsheimer, Robert, 155
Slonimski, Piotr, 136
Soeda, Eichi, 105, 107, 122
Southern, Edwin, 129
Sternberg, Nat, 47
Sulston, John, 81, 128, 129
Sutherland, Grant, 69, 71, 72, 112

Tooze, John, 148
Travaglini, Giorgio, 91
Tsui, Lap-Chee, 65

Venter, Graig, 83, 86, 87, 114, 126
Vickers, Tony, 126
Vogel, Friedrich, 139

Wada, Akiyoshi, 107, 117, 121
Ward, David, 159, 160
Warren, Steve, 68, 69, 70
Waterston, Bob, 81, 129
Watson, James D, 1, 12, 19, 88, 112, 146, 147, 163
Weissenbach, Jean, 24, 28, 131, 134
Wexler, Nancy, 146
White, Ray, 112
Wyngaarden, James, 148

Yesley, Michael, 171

Formaté typographiquement par DESK,
Laval – Tél. (16) 43 68 13 67

Achevé d'imprimer par Corlet, Imprimeur, S.A.
14110 Condé-sur-Noireau (France)
N° d'Imprimeur : 9673 - Dépôt légal : juin 1993
Imprimé en C.E.E.